Jerome Teelucksingh

Caribbean Football Victory

Jerome Teelucksingh

Caribbean Football Victory

Surviving in Paradise

JustFiction Edition

Imprint
Any brand names and product names mentioned in this book are subject to trademark, brand or patent protection and are trademarks or registered trademarks of their respective holders. The use of brand names, product names, common names, trade names, product descriptions etc. even without a particular marking in this work is in no way to be construed to mean that such names may be regarded as unrestricted in respect of trademark and brand protection legislation and could thus be used by anyone.

Cover image: www.ingimage.com

Publisher:
JustFiction! Edition
is a trademark of
Dodo Books Indian Ocean Ltd. and OmniScriptum S.R.L publishing group

120 High Road, East Finchley, London, N2 9ED, United Kingdom
Str. Armeneasca 28/1, office 1, Chisinau MD-2012, Republic of Moldova, Europe
Printed at: see last page
ISBN: 978-620-6-74071-1

Dedicated to Caribbean people at home and abroad

Table of Contents

Chapter 1

Temporary defeats

The small object began to move. It was on top of a large garbage bag in a rectangular area designated for the community's garbage. The area was blocked off by three layers of red bricks which were held together by thin layers of mortar. There was a horrible stench of decaying food amidst the empty tins, broken bottles and cardboard boxes. Flies buzzed near the object. The object made a noise. This was something unusual in Beetham in Trinidad and Tobago, in the Caribbean.

Syl and Derick were nearby playing with spinning tops and toy cars. They were seven years old and this was the first day of their Easter holidays. They wore khaki pants and discoloured jerseys. Their hair was uncombed. Both heard the noise and stopped playing. The cautiously approached the pile. Both boys jumped when they heard the shrill cry. Syl thought it was an evil folklore creature. He began to sweat. Derick was afraid and wanted to run to his home.

Syl saw a small, moving hand from a partially wrapped bundle. 'Aye! look…dat is ah hand.' He was stunned. They moved closer. It was a baby resembling a doll that Syl's sister received for her birthday. Derick hid behind Syl. He was peeping to see the hand.

'Look boy…is two tiny hands and a foot,' said Derick, 'dat look like ah baby.' The situation seemed unreal. 'Yuh tink we should go and tell someone?'

The crying intensified. Syl was confused. 'Yeah is ah baby. I eh know what to do. I tink we should carry it from here.' He relutantly lifted the crying baby and held it close to his chest. It was the first time he was holding a baby. The crying continued.

Derick squeezed his nose. 'It smelling rel stink…like sardines and rotten eggs.'

'How you know dat?' Syl laughed. 'Ah feel you eat dat for breakfast. Lissen I will carry home de baby.' Both boys realized the seriousness of the situation.

Syl carefully walked to his home. His mother had seven children and could not care for another child. She gave the baby boy a bath and placed him in a clean set of used clothes. She decided to give the baby to Derick's mother who had four children. On the first Sunday in October, the baby was taken to the church and baptized. He would now be known as Emmanuel. Derick admired the patience of his mother. She would not be flustered when Emmanuel was crying. She would sit in the Morris chair and give the infant sips of lime-bud tea. It was a homemade concoction used to treat gripe and the flu.

On evenings Derick's father would spray Flit as a deterrent to the mosquitoes. His job was cleaning latrines. It was a dirty job but it was the sole source of income for the family. His simple home was constructed from galvanize and pieces of wood. It had no electricity and he would light flambeaux to ensure his children could do their homework. On mornings Derick's mother would wash the cloth diapers and hang them to dry on the clothesline. She never complained of her neverending household duties.

There was a rumour that the baby belonged to Barbara who was seventeen years of age and became depressed when her relationship with Mark ended. She had thin lips and green eyes. She was brown-skinned and considered one of the most beautiful teenagers in Port-of-Spain. Few knew that she was the product of a forbidden relationship. Her great-grandmother was the wife of a Scottish planter and her great- grandfather was a slave.

It was said that after the birth, Barbara told the midwife to dispose of the baby. Another rumour circulating in Beetham was that the mother thought the baby was stillborn and threw it away in the dump. Three days after the baby was discovered, Barbara left and went to begin a new life in Jordan Street, Enterprise. It was a small community in central Trinidad. Nobody in her new neighbourhood knew of her past. Barbara learnt to be a seamstress and on weekends would do babysitting. Most afternoons, she would be seated behind her Singer machine and either sew or repair garments. She had a small Rediffusion radio that played melodies from the United States. During December she would be busy making curtains for the neighbours.

Nobody knew Emmanuel's father. Some said it was Mark who had abandoned his pregnant girlfriend. He was light-skinned and tall. He often boasted that his father was an American soldier who worked at the Chaguaramas military base. He suffered from epilepsy and was unemployed. He was easily recognized and loved to wear a dungaree and jersey. He would spend most of his time in Hell Yard listening to the steelpan music of All Stars.

In 1966, at nineteen years of age, Emmanuel decided to become a Rastafarian. He had brown skin and brown eyes. These eyes reflected confidence, boldness and a purpose in life. He regularly worshipped at the Ethiopian Orthodox Church in Beetham. The building was unpainted and at the entrance were five neatly trimmed crotons with variegated leaves. His poverty-stricken community was located on the outskirts of the city of Port-of-Spain, the capital of Trinidad and Tobago in the Caribbean.

In October 1968 Emmanuel served as a referee in one of the games in the Southwest Football Superleague. The job demanded a high level of fitness as he was constantly moving on the field. After the game he met Elizabeth, of Indian and African descent, and invited her to view his team in the upcoming Freedom Cup games. They met again at the festival of La Divina Pastora in Siparia. Two months after their initial meeting, Emmanuel was married to Elizabeth at the Ethiopian Orthodox Church and they moved into a shack in Beetham. Elizabeth worked as a domestic servant for a Syrian family in Glencoe.

During the weekends, Emmanuel and Elizabeth would regularly visit the Odeon, Ritz and Pyramid. These were the popular cinemas in the former British colony. They also viewed the latest Hollywood movies at Starlite Drive In and Bel Air Drive In cinemas.

Elizabeth's two brothers played for Mahaica United in the Southwest Football Superleague. She and her friends often went to Palo Seco, La Brea and Fyzabad to view the games. Her father was a defender with the Trinidad Amateur Football Association and was part of the overseas tour to play Northern Ireland in May 1952. This was also the year when the Gold Coast footballers, of Africa, toured Britain.

Emmanuel participated in the small goal competition, 'Freedom Cup…No More Borders,' which targeted the Morvant and Laventille communities. This was a football competition which built a community spirit and empowered youths in ghettos and crime-ridden areas in north Trinidad. Emmanuel's two younger brothers, Jameel and Neveal, were members of the Rock City Finest and Victory teams. Emmanuel's team was Jah Unit and recently defeated One Love by two goals to move into the semifinals. Emmanuel was excited. The semifinal was held in the St. Paul Street Multipurpose Complex. In the seventy sixth minute, Emmanuel had his name on the scorer's list. He received a short pass and tapped it home. Despite a goal from Emmanuel, the score was 2-1 as Jah Unit lost to Phase II. The names of some of the teams in the competition were Barcelona, AC Milan and Flamengo. Belonging to neighbourhood teams with European titles was the closest that these disadvantaged youths would get to international football.

After two years in Beetham, Emmanuel became disillusioned with the Orthodox Church and decided to leave. He joined the Boboshanti Tabernacle at Wharf Trace in St. Joseph. He was associated with the Tabernacle but left after one year because he felt the rules were too strict. The Tabernacle was situated within a camping in which women and men lived. Housing was separate for men and women. Emmanuel was told to refer to men and women as Kings and

Empresses. The camping was similar to a commune as everything was shared and free.

There were regular prayers and songs. Emmanuel had to wear special clothes and a turban. Peanuts were prepared and distributed to men of the camping. The men would be on the highway selling the packaged peanuts to the public. Other items purchased by the camping, for sale to the public, were earrings, seals and necklaces.

Any money obtained was divided among the brethren of the camping. Emmanuel soon became bored. The prayers and songs did not help him to realize that inner search for freedom and identity. He was searching for spiritual salvation and realized it was not at the Tabernacle.

Emmanuel wanted to play professional football. He decided to leave the Boboshantis and begin training with the national football team. He was one of the two reserve players on the national football team. After work on evenings, he would meet other players at the training venue which would be the Queen's Park Savannah, Queen's Royal College or Army grounds. The team did not have the advantage of a psychologist, international competition and live-in camps. The coach, Conrad Brathwaite, was a dedicated person.

Emmanuel's brother, Jameel, had played as a defender for Maple Club which was located at Ariapita Avenue. At meetings, members often played card games as poker and bridge. Jameel had been a member of two steelbands- Casablanca and Tripoli. He met persons in the steelbands who had played for the John John Tamboo Band and the Hell Yard Bamboo Band.

Jameel's specialty was the cello pan and he also felt comfortable playing the guitar pan. He had been involved in minor street fights with steelband members of Invaders, Rising Sun and Destination Tokyo. However, he loved playing football more than being part of a steelband. He eventually devoted his full attention to football.

Jameel with his fancy-foot moves had worn the national colours on Trinidad and Tobago's football team in the game against Jamaica at the Queen's Park Oval in 1958. Others in the team were Gerry Alexander and one of the country's best defenders- Tyrone de Labastide. Tyrone had two nicknames- 'Shortie' and 'The Tank'. Jameel, Tyrone and Willie Rodriguez were part of the national team which toured England in 1959. It was probably the first Caribbean football team to tour England. This was a milestone for Jameel who had dropped out of St. Mary's College when he was fifteen years. His decision to leave school was that he did not feel the subjects being offered were interesting.

One evening in October 1966, at the clubhouse, Jameel was shuffling a deck of playing cards. 'Next year is the Pan Am Games in Winnipeg in Canada. Who yuh tink will be on the final eleven?'

Syl looked at his left foot. He replied, 'Ah sure is not me since ah injured meh ankle and missed practice matches for de past two months.'

Jameel dealt six cards to Syl. Members agreed that Syl was among the great defenders of Trinidad and Tobago's football history who included Dick Rodriguez, Selwyn Murren, Jim Lowe, Arnim David, Hugh Mulzac and Gerry Parsons. The defenders of a football team are the forgotten heroes. They rarely scored goals and the accolades usually belonged to the forwards, midfielders or goalkeepers.

In Winnipeg, the Trinidad and Tobago team played against Mexico, Columbia and Argentina. The score was 1-0 in favour of Mexico and a penalty was awarded to Trinidad and Tobago. The skipper, Sedley Joseph, quickly glanced at the faces of his team. All were reluctant to take the decisive kick. Joseph knew the kicking ability of each player. He turned and pointed to Tyrone. The Trinidad and Tobago team supported their captain's choice.

Tyrone began walking to the penalty spot. He slowly walked as a condemned man who did not want to have the dreaded hangman's noose around his neck. It was not only Tyrone who walked to the spot. It was the entire Trinidad and Tobago walking to that penalty spot. It was a slow and painful walk.

In Trinidad and Tobago, thousands of eyes were transfixed on the black and white images on their television screens. It was ten o'clock in the night. There was a full moon in the sky and it resembled a fluorescent football.

Jameel and Emmanuel were watching the game. The image was fuzzy and Emmanuel adjusted the television antennae.

'Right dat better. De screen also jumping a bit. Must be something wrong with de outside antennae. You ever remember daddy telling us about the 1938 World Cup in France?'

Emmanuel shook his head. He missed his father who had recently died. Both men were unaware that 1938 was an important year as Cuba became the first Caribbean nation to qualify for the World Cup. Few persons associate Cuba with football. The country was better known for its sugar, cigars or former leader- Fidel Castro.

'He told meh dat Italy had beat Brazil 2-1. The star was Giuseppe Meazza who scored the winning goal from a penalty in the sixtieth minute…almost to de end of de game!' Jameel pointed to the screen.

Tyrone grabbed the football and spun it around. He stared at it as if it was a crystal ball. Then he gently rested it on the penalty spot and walked away from the ball.

The Mexican goalkeeper paced his area. He seemed confident but was uncertain of the ability of this short defender from Trinidad and Tobago.

Tyrone trotted towards the ball. He intently watched the ball. Trinidad and Tobago trotted towards the ball. There would be no second chances. He glanced at the right corner of the goal. The goalkeeper watched his face. Tyrone's right foot powerfully struck the ball. Trinidad and Tobago had taken the decisive kick.

The goalkeeper with outstretched arms dived to the right and soon realized his error as the ball sailed into the middle of the goal. Trinidad and Tobago had scored! The ball slammed into the back of the net and bounced out of the goal. The ball seemed eager to continue play.

The goalkeeper was sprawled on the ground. He avoided the gaze of his disappointed teammates who were perplexed that this dimunitive Black West Indian scored the penalty. Sedley felt a burden had been lifted from his shoulders. The Trinidad and Tobago footballers ran to congratulate Tyrone. The equalizer energized Trinidad and Tobago. The equalizer was for the team and it was the lifeline needed to save Caribbean football. This would be a milestone signalling the coming of a new messiah for the masses.

The World Cup dream of Trinidad and Tobago began at the Queen's Park Oval on 7 February 1965. It was Sunday and Emmanuel was supposed to be assisting elders in church. Instead he was on the field in Trinidad and Tobago's first-ever World Cup qualifying match.

Before the game began there was a drizzle for ten minutes. Emmanuel viewed it as showers of blessings. He went to the manager's room and signed for his pants, socks and jersey. At the end of the game he and the other players would return these items. He was quiet in the dressing room and not excited as the other ten men who were eager to go on the field. Sir Solomon Hochoy, the country's Governor-General, visited the dressing room. This made the occasion even more special for the players. The Oval was half-filled with enthusiastic fans who felt that Trinidad and Tobago could possibly qualify for the World Cup. This was a fickle crowd who also would not be devastated if the local team was defeated. The team won 4-1 over Suriname.

There was an impromptu celebration in the dressing room. However, the football gods did not favour Trinidad and Tobago in the return leg. Suriname pounded six goals into the back of Trinidad and Tobago's net whilst Emmanuel managed to slip one past Suriname's goalkeeper. In the game against Costa

Rica, Emmanuel again scored the lone goal. He flicked the ball between the goalkeeper's legs and into the goal. However, he was no hero as Costa Rica won 4-1. At Piarco International Airport there were no crowds to greet him and the team. He proved that goals do not make instant heroes. In the return leg, after the final whistle of the referee, Trinidad and Tobago was again on the losing side as the score was 0-1.

In 1973, Emmanuel had two part-time jobs, as a security guard and cashier. He did these jobs to support his family of four and to repay a bank loan. He did not have the time for football and opted to leave the national team. Whilst at work, he would sometimes listen to the games on radio.

On Christmas Day, Jameel visited Emmanuel and his family. They reminisced about their football past. Jameel said, 'We lucky to play football at a professional and national level. Not many people are given the opportunity.'

Emmanuel nodded. 'Yes, is true.' He felt Jameel should not have quit playing football. Jameel needed to rest. During the past two years he had twice broken his ankle and fractured his leg. The desire of representing their country at the World Cup is realized by only a few talented players. The dream of becoming a rich and famous footballer and to play at the World Cup is beyond the reach of most teenagers. Many are either unaware or overlook the uncertainties surrounding a football career. Poor performance or injuries are factors which prematurely end the promising careers of young sportsmen.

'Maple Club has a standard dey have to keep. Dey have to maintain our quality of players. We had people like Ellis 'Puss' Achong, Clifford Roach and Sydney de Coteau.'

Emmanuel replied, 'Doh forgeh Jar Daniel.'

Jameel nodded and scratched his neck. 'Yes. De Club did win League Championship for four straight years from 1960 to 1963. And then from 1967 to 1969.'

The teams in the CONCACAF Zone were competing for a place in the 1974 World Cup finals in West Germany. In one of the qualifying games in Haiti, the referee disallowed four of the five goals which Trinidad and Tobago scored. On four occasions, the joy that was associated with the ball at the back of the net was replaced with disappointment and revulsion. The final score was 2-1 in favour of Haiti. After the game, some of the Trinidad and Tobago players were crying and hugging each other. Other shocked players were lying motionless on the ground. It was the scene of a funeral. Honesty, sportmanship and goodwill in football had suffocated. The Haitian players were celebrating. Blacks had cheated their Black brothers. The Haitian citizens of Port-au-Prince sympathized with the players from Trinidad and Tobago. They lined the city waving at the visiting team which played better football.

Emmanuel switched off the radio. He was in a pensive mood. He thought about the ideals of the grand Haitian Revolution orchestrated by Toussaint L'Ouverture. Many persons admired Haiti as being the first independent Black nation in the Western Hemisphere. The Duvaliers, in the twentieth century, made the Revolution a mockery.

Next day during dinner, Emmanuel was reading *The Evening News*. He was shocked to learn that the football team was stranded in Puerto Rico. This was due to their late arrival in Puerto Rico and missing of the BWIA flight to Trinidad.

'Imagine all de time BWIA planes always leaving late and suddenly dey want to leave on time,' said Emmanuel. He sipped his coffee and then shook his head. 'Dem is de worse.'

His wife nodded. She did not like football and was more interested in hockey. She was captain of Carib Magnolias. The team was ranked third after Maritime Checkers and Malvern. Yesterday at the National Hockey Centre in Tacarigua, she was at the right place to rifle a shot past the goalkeeper. Two penalty strokes in the final game earned her the 'Most Valuable Player' award.

On Christmas Eve in 1973, Emmanuel was among the small crowd of one hundred persons who waited at Piarco International Airport to greet the returning footballers. The football team would win three other games. They defeated Guatemala 1-1, beat the great Mexico by four goals and the Netherlands Antilles 4-1. Trinidad and Tobago would be runners-up to Haiti. Fate had decided that Trinidad and Tobago would not be attending the World Cup in 1974. The fighting spirit of Trinidad and Tobago never departed. It would remain dormant for more than a decade.

Emmanuel decided he would work as a taxi driver transporting tourists from the airport to hotels in Port-of-Spain. He would be earning less but it was more enjoyable than working as a cashier and security guard. He worked from Monday to Saturday during 9am to 2pm. He avoided working in the night. Emmanuel was also a staunch member of the ruling political party United People's Party. He was one of the delegates who regularly attended the party's annual convention.

One day whilst transporting Stephen, a Canadian tourist, to Port-of-Spain, Emmanuel encountered traffic on the highway. This was due to policemen and officers from the Traffic Branch who were checking for drivers without car insurance and heavily tinted car windows.

'Excuse sir, why are those half-naked people running across the highway and chasing after that truck into that compound?' asked Stephen.

'Oh dat place is a major tourist site in de Caribbean. That place is known as the Beetham Landfill Site but it is fondly called de Labasse. Dem residents is diehard supporters of de ruling political party dat always neglecting dem.' Emmanuel wiped the sweat on his brow. There was a small fan attached to the dashboard. He turned the knob to the highest level. He thought about the different accounts of his birth. He was unsure of the truth and recalled one in which he was found amidst the garbage.

The tourist repeated the site's name, 'La Bassé?'

Emmanuel smiled. 'Lissen, doh make it sound like a high-fuleted, big French word. That place is a dump, all de garbage trucks does bring rubbish there and de poor people running to salvage stuff. Once ah did go dere to look for ole wood and galvanize to repair meh house.'

'But that is hazardous to their health. Look some of them are even barefeet. They could contract deadly diseases and chemicals would be absorbed into their bodies.'

Emmanuel paused. He was not sure if it was a health risk. 'Not for hell dem go lissen to de government who want to prevent accidents. De way I see it, is better dey do dat than go and tief in people house or murder.'

'Do you really think so?' asked Stephen.

'Yeah.'

Emmanuel stopped and allowed two persons to cross the highway to return to the Beetham. Both persons were carrying rusty pieces of iron and a black plastic bag. Some of the youths of Beetham would use the short pieces of iron as goalposts.

'Dem is garbage peong. Dey not only looking for food but material for dem ragadang cooboose.'

Stephen nodded. 'By the way, what is your name?'

'I man furse name is Emmanuel buh people does call me Locksman.'

Stephen liked the alias better than the driver's real name. 'So Locksman how does the Beetham deal with this waste?'

'Some is buried and the rest burnt.'

Emmanuel looked at the houses on the Beetham Estate. He saw five boys playing football in the yard of one dilapidated house. They were bareback and two had dreadlocks. Broken bricks and garbage bags were used as the goalposts. Their football was soft and shredded. Emmanuel remembered seeing a documentary about poverty-stricken boys in an African village who were kicking a ball made of rubber bands and old rags. The boys on the Beetham

Estate reminded him of the boys in the African village. Both groups were carefree and laughing. They seemed unconcerned about the present and their future. Life was fun. These were boys in the Beetham Estate who should have been at school. The government did not care and their major concern was securing the votes of the boys' parents. Emmanuel wondered if they were too poor to attend school or did not appreciate the value of education. He wanted to go and encourage them to return to school but he had a passenger to transport to the city.

'So you ever tasted Trinidad rum?'

'No. I would like to try it.'

Emmanuel glanced in the side mirror and then swung to the shoulder of the highway. He placed his hand under the seat and pulled a lever. The trunk popped open. 'Come out for a drink.'

Stephen looked confused. He alighted from the car but did not see any bars or restaurants. Emmanuel opened the trunk and Stephen cautiously moved towards it.

Emmanuel opened a cooler and chose a small bottle of rum. 'Here try this. Is one of my favourites.' He decided on a beer.

Stephen took the bottle and had a worried expression. 'Is it against the law to drink in public?'

'No. Not here in Trinidad. Having some beers and rum in your car trunk is jus' as important as de spare tyre and insurance.'

'What about drinking and driving? Suppose de police spot us and stop?'

'Very drunk people should not be allowed to drive.' Emmanuel smiled. 'If de police stop then all we do is offer dem a drink.'

Stephen took a mouthful of the rum. He enjoyed the sweetness. He observed the nearby flora. He wore a hat, polka dot shirt, black pants and sandals.

After ten minutes both men returned to the car and continued the journey.

As they entered Port-of-Spain, Emmanuel spotted his familiar eating outlet. 'Look, ah roti shop, leh we buy a roti. Yuh ever eat roti?'

'No, and I would like to try one.'

'Dis roti shop does sell nice chicken and potato roti, it does taste good. Is meh brudder favourite.' Emmanuel glanced at the tourist to observe his reaction. 'Is not like odder shops where you could swear is dog meat or corbeau yuh eating. Next door selling pastries like cassava pone. De owner have plenty face to wash and is a money satan.'

Emmanuel parked the car near the pavement and both men entered the small shop.

'Yes, morning Locksman,' said Derick who was wiping the counter. 'Can I help you?' He picked his nose.

'Morning. Give me two chicken roti and two red Solo.' Emmanuel withdrew his wallet from the back pocket of his pants. He opened it.

'Don't bother I will pay for it. My currency is worth more.' Stephen looked at the owner of the shop and asked, 'How much is it?'

Emmanuel nodded and waved his hand.

'Five U.S or Canadian dollars will cover everything.' Derick then looked at Emmanuel and smiled. 'Yuh stop buying shrimp roti?'

Derick played bass steelpans for Desperadoes. He joined when this steelband won the Panorama title in 1970. Derick admired two steelpan arrangers- Lennox Mohammed and Jit Samaroo.

Emmanuel shook his head. He returned the wallet to his pocket. He took a toothpick from a glass bottle on the counter.

Derick waved his hand. 'Doh bother with me. Ah kicksin.'

The tourist looked at Derick. 'Last year I read an article about marine pollution. There is now aluminum, mercury, lead and zinc in seafood.'

'Fuh true?' asked Emmanuel. 'What doh kill yuh does make yuh stronger.' He cleaned his teeth with the toothpick then threw it on ground.

'Yes. The raw sewage and fertilizers are washed down by the rain into the oceans. I living in the USA and dat is why I stop eating at Long John Silver and Red Lobster.'

Derick placed chicken, channa and potato in the middle of the roti skin. 'Any pepper Mr. Tourist?'

'A little for me,' said Stephen. He keenly watched Derick fold the dhalpuri skin. It was a new experience for him.

Both men sat down and began to eat the roti. The tourist began to greedily consume the roti and pieces fell on the plate and floor.

Derick laughed. 'Yeah take it easy. Yuh have time for ah next one when yuh finish.'

Stephen stopped chewing and pointed at the roti. 'This is ah great invention. Who invented the roti?'

Derick heard the question and offered an answer. 'We not sure. Buh fuh sure is not ah American, Chinese or European person.'

That answer seemed satisfactory for Stephen who continued chewing. He mumbled, 'I always thought hamburgers and hot dogs were the greatest.'

Emmanuel was quiet. He was concerned that the cost of a bottle of cooking oil had increased by seven cents. His wife was a member of the Housewives Association of Trinidad and Tobago and planned to participate in a protest over the price increase. He opened the newspaper and browsed at the advertisements. There was the Texaco Giant Slogan Contest with prizes including a RCA television, Zenith radio and GE refrigerator. His home needed a new refrigerator and radio. There was another competition - 'Tickets for the Test Competition' in which an entry form would be given with the purchase of two dollars worth of gasoline at any Shell service station.

'So Locksman yuh brudder still selling caca poule rum?'

'Yeah. He is ah lucky fella. Babylon nearly ketch him at a whe-whe game in Rio Claro yesterday.' Emmanuel stopped chewing and swallowed. 'When dat mark buss, babylon go buss he tail. Yuh know he having some problems again with he queen.'

Derick wiped the counter. 'He have to be like me- with no chick nor chile.'

'Meh brudder have to get some commanding powder or sweat rice or take a good bush bath. Before he did married she, ah tired tell him that she is no good. She is a real canal conchs.'

'Ah dat roti was good.' Stephen licked his plate and three fingers. Then he drank the Solo and requested another roti. Within five minutes he consumed the second roti.

'Come let's go now, and buy some pone next door.' Emmanuel watched the clock near the window. 'And den ah go drop yuh off at Holiday Inn. After ah have to buy some groceries for meh lady.'

'Later Derick we go talk later in de week.'

'Yeah later Locksman, say hello to de madam.'

The bakery was crowded with persons buying hops bread. The cashier was busy and the two attendants appeared flustered.

Stephen took a bag with pastries and placed it in the side pocket of his suitcase. Emmanuel parked near the hotel and carried Stephen's suitcases to the receptionist's desk. Stephen quickly paid Emmanuel and bid him farewell.

Emmanuel departed for Beetham but after fifteen minutes realized that Stephen had forgotten his camera on the front seat. Emmanuel examined it. The brand was Kodak and it looked expensive. He returned to the hotel and was surprised to see Stephen's luggage still near the counter. He approached the receptionist. She had a scarf and wore a green skirt and blue blouse. The colour of her earrings matched her blouse.

'I want to return dis camera to a tourist ah juss drop off here. He was wearing a hat and polka dot shirt.'

The receptionist smiled. 'Yes I remember you. Leave it here, he should be here shortly.' She pointed to a door in the corridor. 'He is in the toilet.'

Emmanuel gave her the camera and began to leave.

'Hey Locksman! I thought you had gone.'

Emmanuel turned around and saw Stephen walking towards him. 'Yuh left yuh camera and ah returned it at de desk.'

'Thanks my good man. You people are very honest.' Stephen had a hand over his stomach. 'I was in the toilet…that was some real hot pepper!'

Emmanuel looked at the clerk and then at Stephen. 'Nah man. Dat is de airline food. Dem ting does be stale and spoil.'

'Hey Emmanuel! shouted Philip. 'You here today?'

Emmanuel spun around and saw his neighbour, Philip, who recently returned from England. Philip, at nine years of age, migrated with his parents to London. He worked for twenty-four years as a clerk in the House of Commons in London.

'Yes, you want a lift back home? I heading that way now.'

Philip nodded and sat in the passenger seat. They spoke about football in the Caribbean. Philip exited the car and said, 'Come over let's talk some more.' Emmanuel checked his wristwatch and agreed.

Emmanuel knocked and entered his neighbour's home. Philip was in a hammock and leisurely smoking a cigarette. 'Sit down I have organge juice, beer and cake. Listen mate, I was always supporting football in England especially the FA Cup. I'm a supporter of Everton but used to support Leeds United. There were always hooligans in some of these games and sometimes they fought after games. I've met a few rowdy supporters from West Ham United and Liverpool. But me and my mates took precautions from those buggers.'

'What does de police do?' asked Emmanuel.

'Bobbies…the police, are always present at the games but it's difficult to control the fans. Once I saw a few fans fighting in the loo. Luckily I had my brolly to defend meself just in case the buggers attacked me.'

Emmanuel frowned. He did not like to hear a Trinidadian using foreign terms as brolly and loo. 'Have you ever met any famous footballers?'

'Of course, I have the autograph of Cliff Bastin who scored more than one hundred goals for Arsenal and I also got a picture with me and George Best. They no longer play football. I was there to see de World Cup Final in 1966.' He gently tapped the cigarette and the burnt ashes fell to the ground. 'Mate, let me tell you that footballers like most sportsmen have a short shelf-life. Some retire due to injuries, age or because of personal problems.'

Emmanuel nodded. 'Yeah I agree. In Trinidad is de same thing.'

Philip looked at the cigarette butts on the ground. 'Mate, do you know some areas even have football games for fun? I've seen an annual mud-football game at the Leigh-on-Sea regatta and there is one at Ashbourne in Derbyshire.'

'Yeah, here it does have a similar ting. Every year is fete match in cricket and football. On Christmas Day in Santa Cruz it does have a nightie and pyjama football game.' Emmanuel sipped a glass of juice and cleared his throat. 'Have you seen any of de visiting foreign teams play football whilst in England?'

'Of course mate. One game that I always remember was in 1950. A team from Nigeria came to play a match against Bishop Auckland. And, believe it or not, these lads from Africa did not wear boots. They wore support straps on their feet. They played some good football. One thing about the British is that everybody likes to play or watch football. One of my mates who died recently used to play football when he was a schoolboy in the 1930s. He would always tell me about some game where he played for Walthamstow and they defeated the chaps at West Ham. I remember as a boy seeing long lines at football games with mothers with babies in prams. They were waiting to buy tickets. In some areas bales of straw were put on the football ground before a game to protect it from the frost.'

'Women play football in England?' Emmanuel asked. He did not know the meaning of mates and chaps.

'Yes, there was one group me vaguely remember...something like Maldon Ladies Football Club. When I was a boy, I heard stories about Miss Casey and Miss Millers from Bradford who was the football coach for their schools.'

'Why yuh tink dem English people could play better than we in the Caribbean?'

Philip threw the final piece of cigarette in the drain. 'Well for one thing, being a footballer is a full-time profession in England. Those footballers get paid big money. I once saw some injured footballers getting a high pressure shower massage. Them English coaches take the training seriously and some had giant sling-shots to shoot the ball in the air. They also now use automatic crossing machines. I saw it being used.' He coughed and cleared his throat. 'I saw Wolverhampton Wanderers using it.'

Emmanuel asked, 'You did use to play football in England?' He looked at a keskidee on the fence. It had black, yellow and white markings on its wings and body. The bird flew off and landed on the branch of a willow tree.

'Not for a team. Me brother used to play centre-half. But he was never serious about football. Imagine he used to enjoy being a ticket tout outside the stadium rather than being a footballer in the stadium! My dad use to talk of his days as a footballer but me don't believe him. He was a bum, always betting on horses and football. He won five thousand pounds in the Littlewood pools and after two months lost it all in gambling.' Philip paused. He wanted to provide an accurate picture of his father. 'He had all this talent but wasted it on women and gambling. He was like that character from the comic strip- Andy Capp. Me mother was the one who really helped me and me brothers and sister geh through school. She was a hard-working nurse.'

Emmanuel smiled. 'So who do you believe was the greatest footballer England ever produced?'

Philip lit another cigarette. 'It's difficult to decide mate. My personal opinion is that there were two greatest footballers- Bobby Moore a defender and the goalkeeper- Gordon Banks. Did you see the famous save by Banks in the 1970 World Cup?'

'Yes, of course. Pele had gotten a cross and jumped to head the ball inside the goalpost. Pele ran off to celebrate and did not wait to see the ball enter the goal. But, Banks stretched and with one hand pushed the ball over de bar. So is it true that plastic footballs are stronger and better dan de leather ones?' Emmanuel was planning to buy a football for his son as a Christmas gift. The plastic ones were cheaper than the leather footballs.

'The plastic might be stronger but footballers all over England and Europe feel it more comfortable kicking the leather ones. Some footballers say that the ball moves better in the air.' He threw the cigarette butt on the ground and stepped on it.

'Listen, it was nice to hear 'bout England and football. Ah want to go home early today.'

Philip nodded. 'Yes mate, thanks for the company.' The keskidee returned to the fence. It had a small grasshopper in its mouth.

Emmanuel began walking to his home. He checked his wristwatch. He had to reach in ten minutes if he wanted to see the beginning of the nightly news.

On Sunday Emmanuel decided he would not attend church. He and Philip went to play football in the eastern section of the Queen's Park Savannah. He enjoyed playing and after two hours returned home.

Emmanuel returned home and met his wife who was busy preparing lunch. 'Dat Sunday morning sweat is real good. It had odder players from behind de bridge.'

Elizabeth steupsed. 'Lissen, doh make it ah habit. Sunday is ah holy day and is not to kick ball.'

The next evening Emmanuel went to visit a Boboshanti priest residing in Morvant. He had an appointment to meet the priest at three o'clock. The priest wore an Adidas track suit and had a black headwrap. His red and green jersey had a small picture of a golden lion. He lived alone in a simple, unpainted house. Four louvres were missing from the windows. At the back of his home was a tall poui tree with brilliant yellow flowers.

Emmanuel was offered a seat in the living room. He nodded and sat down.

The priest said, 'I met you when you were a member of the Boboshanti Tabernacle. You have a pure heart.'

Emmanuel did not remember meeting the priest.

'I'll return in a short time.' The priest entered a nearby bedroom. He returned with an object wrapped in gold and green cloth.

It was carefully placed on a wooden table and slowly unwrapped by the priest. The object lay exposed on the table. It was an old Bible and in pristine condition.

The priest told Emmanuel, 'This was given to me by a man whose great-grandmother was brought from Africa as a slave. She used to live in Corbeaux Town. Come and see it.'

Emmanuel was amazed. He carefully turned the pages. It was a King James version and was undated. The Bible had illustrations of Old Testament and New Testament characters. All the characters were dark-skinned and of African descent. He wondered if this was the oldest Bible in the Caribbean.

'I want to give you dis Bible for safe-keeping. You will soon have use for it.'

Emmanuel reluctantly accepted it and thanked him. He returned to his car and headed for his home.

Emmanuel never liked the police. They would regularly visit his home to search for stolen items. If there was robbery at Valsayn or Westmoorings, the police would go to Beetham and randomly search a few homes.

On Easter Sunday the police visited Emmanuel's home. A policeman banged on the door.

Emmanuel moved the curtain and peered through the window. He was not surprised to see two policemen carrying rifles. On their waist were holsters with loaded guns.

One was skinny and had a nose which made him resemble an eagle. Last week he had searched two homes in Emmanuel's neighbourhood. The other policeman was skinny and similar to a walking corpse.

Emmanuel opened the door. 'Hello,' he mumbled. He stared at the potted plants near the doorway and did not look at the faces of the officers. 'You all come to search again?'

'Yuh tink we born yesterday? We know you is involved in de robbery of de home in Westmoorings and have stolen goods.'

Emmanuel remained calm and moved aside to allow the officers to enter his home. He adjusted the mat and watered the potted plants. It was humid and he looked at the sky. There were dark clouds and he hoped it would rain.

The impertinent officers entered the kitchen and began to open and slam cupboards. They emptied drawers containing utensils. Strips of sticky yellow flypaper hung from the ceiling. There were dead flies, moths, mosquitoes and a small lizard on the strips.

Both officers also ransacked the living and dining rooms. Raphael was upstairs and heard the commotion. He was frightened and knew it was the police. He quietly hid in the bathroom.

'Yuh want to search upstairs?' asked Emmanuel.

'Only if yuh woman up dere! Ah want to search her,' replied the gaunt officer.

Emmanuel acted as if he did not hear the police officer's comment. He adjusted one of the sleeves of his long-sleeved shirt. The shirt was not ironed and resembled an old flour bag.

The police did not search upstairs. They left empty-handed.

Raphael slowly descended the stairs. 'Daddy de police gone?'

'Yeah come down and relax. Doh tell yuh mudder nothing. She go only geh vex and worry 'bout we safety.'

Raphael went into the living room and switched on the television. The Cosby Show was airing. He liked the show since it depicted positive family values among Blacks.

Emmanuel went into the kitchen. He organized wares in the cupboard and disposed of two broken dishes and three teacups. He saw a piece of paper and pencil on the kitchen counter. He wrote a short note- 'Buy ham tomorrow'.

Elizabeth returned from work and began cooking dinner. Emmanuel entered the kitchen and looked at her. He sensed that she was in a bad mood.

'You always complaining 'bout how I does look. If is not meh belly, then is meh flabby arms and big bottom. Is you who want me to get botox injections.'

Emmanuel was ready to reply. He opened a pack of orange juice and placed it in the refrigerator. 'Lissen, ent ah tell yuh dat…ah like yuh as yuh is. Buh clean up yuh face…yuh have a big set ah moustache and beard. Go pluck dem hairs or someting.'

Emmanuel left before she could reply. He went into the study room and closed the door. He sat on the swivel chair and looked for a good book to read. There were two shelves with magazines and journals and four shelves with fiction and History books.

Raphael was upstairs. He was dressing to go to the movies with friends. He was struggling to get a contact lens in his left eye. Later that night when he returned home, he could not remove the lens from his right eye. His parents and neighbours unsuccessfully tried to remove it. Eventually, he had to be taken to the local hospital where a doctor provided an injection for the eye's muscles to relax. At 2am the lens was finally removed.

Emmanuel told Raphael, 'Yuh ever see Rastas wearing glasses and contact lens? Real rastas doh wear glasses. Dat is for de bourgeoisie Rastas…de fake Rasta.'

Raphael asked, 'Suppose ah not seeing good?'

Emmanuel smiled. 'Son, pull de sides of yuh eyes so it could be like chinee and yuh go see better. Do whatever buh try and doh wear glasses.'

The next day Elizabeth decided to remove the facial hair from above her upper lip and chin. She went to a beauty saloon in the neighbourhood. The waxing was not a costly procedure.

Elizabeth's neighbour, Ann Williams, was a feminist who was a part-time lecturer in Gender Studies at Thames Valley University in England. Ann was spending two months holidaying in Trinidad and Tobago.

Ann commented on Elizabeth's new look. 'Gyul yuh should not have gotten rid of de facial hair. All we sistahs need to be independent and take charge of we body and we life.'

Elizabeth had a serious expression. 'Ah wanted to try a new look. Doh worry nothing else of me will change.'

Both women continued talking for ten minutes then Elizabeth left to cook dinner.

That evening during dinner, Elizabeth raised the topic with Emmanuel. 'Ann feels dat ah should not have gotten rid of de hair on my face.'

Emmanuel steupsed. He stared at his wife. 'She saying dat cause it not on she face. Ent is she say dat all dem moles on yuh face looking good and not to remove it?'

'Yes. Buh she tell me 'bout natural beauty.'

'Doh worry with Ann. Wha' she know 'bout natural beauty? She bleach she skin and straighten and colour she hair. And now she hair looking thin and scanty. After doing all dat she still cyah pick up a man in England or here. Even if she use sweat rice or lef' han' dumplin' she cyah get a man. She will always be short, ugly and manless. You ent see something wrong with she mind.'

Emmanuel hugged his wife and kissed her forehead. 'Now yuh looking like de woman ah marry. Go and sleep and ah will close up downstairs.'

Next day Emmanuel returned home with a smoked ham. It had a label with the word- Piggy. It was purchased at Hong Wing's Grocery on St. Vincent Street. It was in the corner of the grocery for almost two years. He also purchased a Butterball turkey at Tru-Valu grocery and a football from Sports and Games. The football was a Christmas gift for Raphael.

Elizabeth was furious when she saw the purchase. 'More ham again? The freezer have two hams to be baked! De fridge have sliced ham for sandwiches. De cupboard have tinned ham. Look we have no milk or carrots in de house. Anytime ah give you a list of items to buy you does go to de grocery and market and bring back ham or anything you feel to eat.'

'Ham is important for we to grow. December is de month for ham.'

'Yes, to grow big and fat like a pig! Eat all de damn ham yuh want and somebody go mention it in yuh eulogy.' Elizabeth paused. 'Lemme stop wasting my breath because ah know next week yuh going to buy more ham.'

Emmanuel found a pencil on the table and wrote a short note. He stuck it on the door of the refrigerator. The note had – 'buy less ham and coffee.' He underlined the word ham.

On Christmas Eve in December 1980, three police jeeps entered the compound of Emmanuel's home. Two policemen kicked down the door. A policeman shouted, 'Where de Rasta man? Speak woman or ah go shoot yuh!'

'Wha' happen? Meh husband innocent.'

'Where de Rasta who live here?' Another policeman shoved a gun in her face. 'Woman talk fast. Yuh want a bullet too? Where de damn Rasta?'

She stood quiet. The officer cocked the gun. She pointed to the next room.

The policemen entered the room. They saw Emmanuel sleeping on a bed. Three shots echoed throughout the house. Elizabeth began to cry. The policemen left the room and jumped into their cars and sped away. Curious neighbours rushed to the house.

'Emmanuel? Emmanuel?'

There was no response from the room.

She entered the room and fell on her knees. His lifeless body was on the ground. From under his body, a pool of blood began to slowly expand. She began to bawl and clutched her chest and shouted, 'De DAMN POLICE KILL EMMANUEL IN COLD BLOOD!' She looked heavenwards and shouted, 'GOD WHY YUH DIDN'T DO SOMETING? GOD WE IS PEOPLE TOO!'

Raphael was twenty years old when his father died. He was playing football on the street. He heard the news from a neighbour and quickly returned home but the police had already departed. The shocked son cradled his dead father and loudly cursed the police and the government. That evening, an angry Raphael took a bottle of pitch oil and poured it over the blood-soaked bed sheets. He lit a match and threw it on the mattress. He cursed and swore that he would never trust the police and would focus on removing the ruling political party which claimed to represent Blacks. His mother warned him to leave vengeance to the Lord.

Two days later, Elizabeth saw a note on the refrigerator- 'buy less ham and coffee.' She smiled and immediately recognized Emmanuel's crooked handwriting. It was a traumatic time and she borrowed money from a cousin and the Eastern Credit Union to pay for the casket. The funeral was held at the Ethiopian Orthodox Church in Beetham. There were members from groups such as Bobo Shanti House, African Unity, Twelve Tribes of Israel, Universal Love, Nyabinghi and Rastafarian Corporation. Neveal delivered the eulogy. He read two verses from the Bible that once belonged to the Boboshanti priest. Jameel and Derick were among the pallbearers. Both men had tears in their eyes. Elizabeth fainted during the singing of the final hymn. Raphael told the congregation that the police and the politicians were part of Babylon.

The murder of Emmanuel was neither reported on the radio nor the two daily newspapers. It seemed that nobody in Trinidad and Tobago and the world cared. It was just another dead Rasta. Three months after the murder, a grieving Elizabeth died of a heart attack.

After Emmanuel's murder, Philip became frightened and purchased a one-way ticket to England. He decided to spend the rest of his life in England. Philip also warned West Indian immigrants, in London, to avoid visiting Trinidad.

It had been nine years since Emmanuel was murdered by the police. Raphael had left Beetham and was squatting on Freedom Street in Enterprise in central Trinidad. The tragedy still haunted him during the nights. He was twenty nine years old and a Rastafarian. Nobody was ever charged or convicted for his father's murder. Raphael would regularly remember the Biblical quotation repeated by his mother after the murder- 'Vengenace is mine says the Lord.' It made him more tolerant of the injustice.

On weekends Raphael worked at Pap's Poultry Depot. He would weigh, pluck and cut chickens for customers. After weighing the chicken, he would place the chicken headfirst into a funnel and cut its neck. After a few minutes the motionless chicken was thrown into an electric plucking machine. This was a steel cylinder with rubber spikes attached to the inside. Water was thrown into the plucking machine whilst the chicken was being tossed about in it. After three minutes in the plucking machine, the featherless bird was removed and washed in a nearby sink. The bird's head was cut off and the intestines were removed. These were thrown into a pile on the ground. The headless body would be cut into quarters or smaller pieces. The cutting was done on a piece of tree stump. Some persons would order ducks, broilers or layers which were more expensive than chickens. The duck was a delicacy and this was usually roasted to add flavor. The roasting was done by applying an open flame to the plucked duck.

The drains nearby Pap's Poultry Depot were stagnant and covered in a layer of feathers. Blackbirds pecked at the intestines that were washed into the drain. Flies were everywhere. Raphael began to despise killing the chickens. After seven months he left the job and decided to become a vegetarian.

Chapter 2
Almost victorious

Raphael was determined to change society and volunteered to be a foot soldier for the newly formed National Construction Party in 1986. He mobilized voters in Longdenville, Enterprise, Endeavour and Cunupia. In 1986, the calypsoes 'Sinking Ship' and 'Vote Them Out' were the campaign songs that were popular throughout the country. The huge turnout at the pre-election rally at Woodford Square was the sign that the National Construction Party would be the eventual winner at the polls. It was a pyrhhic victory. At last there would be a new political party to reclaim the country's lost paradise.

Raphael was excited that in three days would be the CONCACAF Zone qualifier between the United States and Trinidad and Tobago. His father had often told him about the chance that Trinidad and Tobago had in 1973 to qualify for the World Cup. His father had seen the match on Trinidad and Tobago Television.

On Friday 17 November 1989 there was excitement throughout the country. Some persons had fetes and parties. The celebrations had begun. Raphael decided to go with his friends to witness this historic game at the National Stadium. Special 'Football Massive' screens were placed in Skinner Park in San Fernando, Chaguanas Junior Secondary School and the Queen's Park Oval. Trinidad and Tobago and Costa Rica were at the top of the standings. The other teams were United States, Guatemala and El Salvador. Trinidad and Tobago hoped to become the twenty-fourth and final nation to qualify for the 1990 World Cup soccer finals in Italy.

A fortnight ago Raphael and his friends purchased three tickets to sit in the covered stands. They arrived at 2pm and the stadium was already filled even though the game would begin at 3.30pm. There were cultural activities to entertain the massive crowd of 41,000 persons who were part of the nation's 'Road to Italy' campaign. It was a 'Day of Red' as persons wore red jerseys, pants and bandanas. Others wore jerseys with numbers similar to that worn by the footballers of the Trinidad and Tobago team. Some waved national flags whilst others had their faces painted in red, white and black. Drumming emanated from around the stadium.

During the game there were vendors selling a variety of snacks and drinks. One vendor had coconut water in plastic bottles.

'Mayaro Fresh. Coconut water! Get your coconut water!'

'I'll take three bottles,' shouted Ravi. He waved a twenty dollar bill. He took the bottle and began to read the label. 'Buh dis is ah new product. Lawd is five dollars dey charging meh for a lil bit of coconut water!' He looked at the vendor. 'Like dese coconuts come from China?'

The vendor smiled and took the picong in good spirit. He gave Ravi five dollars and departed.

Trinidad and Tobago's players had red jerseys and black pants. They were known as the Strike Squad and when they walked onto the field the crowd erupted in applause.

The referee's whistle initiated the battle of the gladiators. The left winger, Leonson Lewis, passed two defenders and made a cross for Russell Latapy but the ball was intercepted by Tab Ramos of the United States.

The ball was in the United States half and Paul Caligiuri roughly tackled Dwight Yorke. The referee blew his whistle and awarded a free kick to Trinidad and Tobago. The crowd booed at the player who committed the foul against Dwight. Both Dwight and Russell were closely marked by United States defenders.

'I feel that we have a chance to qualify for Italy because Hugo Perez not in de game,' said Surendra. He began biting his fingernails.

Raphael nodded. 'Yeah dat is true talk. He is a key player. Buh we have de Little Magician to help we. He good at spraying passes all over the field. All we need is a draw to qualify but de U.S. in a must-win situation.'

Clayton Morris blasted a shot at the goal of his opponents. The stadium erupted in cheering. Tony Meola, the goalkeeper for the United States, dived to his left and deflected the ball away from the goal. Some spectators were on their feet. All eyes were on the ball. A country's dreams and hopes rested on that one ball.

Ravi was uninterested with the game. He was busy eating a hot dog and drinking coconut water. After completing the hot dog he took a hamburger and began to greedily devour it. He would occasionally glance at the game but was more concerned with his food.

During the first half, the fans were eagerly following every move of their favourite players. Surendra was hoping that a goal could be scored by Dexter Francis, Kerry Jameson or Marvin Faustin. Dexter and Leonson were tightly marked by two players from the United States. Raphael believed that Brian Williams, sporting dreadlocks, was a solid defender who would frustrate any goal scoring opportunities of the United States. Kerry sent a long throw from the sideline to Dwight who trapped the ball and ran towards the goal. He received a rough tackle from a defender and fell on the ground. The crowd vocally expressed their disappointment.

Ravi completed the hamburger and strained his neck searching for another food vendor. Near to his feet were four empty Carib cups, cigarette butts and two empty coconut bottles. He looked at the persons standing and seated in the aisles. It was obvious that the stadium was overcrowded. He was worried about his safety but did not inform his friends.

Twenty five minutes had passed. It was a tense game. Both teams were evenly matched. Some persons were checking their wristwatches. Soon it would be half-time. Trinidad and Tobago's goalkeeper waved and shouted at a defender.

Ravi left his seat and returned with two packs of Broadway cigarettes, a large cup of Coca-Cola and a box of Kentucky Fried Chicken. He opened the box and licked his lips. He bit into a drumstick that was covered in ketchup and mustard. He stopped chewing and took two gulps of his drink.

A teenager passed near the group of friends. 'Walls! Flavorite! Walls! Flavorite!'

Ravi's mouth was filled with food and he quickly raised his arm. His began to wave his hand. He was as eager as a student who knew the answer to a difficult question.

The vendor spotted him and approached the group. 'Wha' yuh want? Walls? Flavorite?'

Ravi swallowed and looked at Surendra and Raphael. 'Allyuh want anything? Ah feeling lickrish.'

Raphael and Surendra smiled and shook their heads. They were absorbed in the game.

'Give me a Walls and a vanilla Flavorite.' Ravi stuck a messy hand in his shirt pocket and gave a ten dollar bill to the boy. 'Keep the change. Yuh have any Cornetto cones?'

The vendor checked his container. 'No. Buh it have a woman selling some by de scoreboard.'

'If yuh bounce her up, tell her to pass dis way na. Good boy.'

'Ok.' The vendor continued walking and began his familiar chant. 'Walls, Flavorite! Walls! Flavorite!'

Ravi began to chew faster. He wanted to complete the meal so he could eat the two ice cream cones which were quickly melting.

Surendra jokingly advised, 'Aye doh eat like de food running away or is yuh larse meal. Take yuh time, nobody will take yuh food.'

Ravi smiled and continued chewing. He found it was difficult to eat and view the game.

In the thirty-first minute, the United States scored a goal. The sea of jubilant, red supporters suddenly ebbed away. The few supporters of the Americans waved their flags and clapped.

The goal was a mystery. Paul Caligiuri fired a powerful shot at the goal. It was a left-footer about twenty metres from the goal. The airborne ball curled and dipped. The goalkeeper, Michael Maurice looked upwards and was partially blinded by the sun. Also, the Trinidad and Tobago defenders who were near Caligiuri, partly blocked the view of Maurice. The ball struck the back of the net. This meant the hopes of an entire nation were at the back of that net. The net was a hangman's noose which was now around the neck of a nation.

Raphael stared at the ball bouncing between the goalposts. He felt weak. The ball reminded him of a helpless fish caught in a net. It was a thrashing fish that would never be free, a fish destined for the cooking pot.

An enthusiastic Tab Ramos grabbed the ball from the net and carried it to the half-line. The player was not merely carrying a ball. The United States player was taking away the promise of the Strike Squad and the dreams of a country. Michael Maurice had a disappointed look. History would not be kind to him.

Raphael looked at the diverse crowd. He joined others to complete the Mexican wave which was circling the stadium. He looked at the ball being kicked around.

Men and women of different ethnic groups, cultures, classes and religions were united. One ball and eleven men had united a country. One football game did more than all the politicians in uniting Trinidad and Tobago.

During the interval, Surendra decided to buy corn soup. He was distressed that the score was 1-0. The atmosphere was tense. Surendra could sense the change. The rhythm of the drums had changed. He slowly passed the long line at the KFC outlet and saw Ravi at the counter.

Ravi clutched a slice of pizza and two hamburgers. There was a beer bottle in his back pocket. He shouted at the serving assistant, 'Look na, do meh favour and give me a big thigh or drumstick. I eh want no wing. Yuh know ah is ah meat mouth.' He paid the cashier and moved to the side where he began to squirt ketchup and pepper on his fries and chicken. He departed and looked for a vendor selling hotdogs.

Surendra had a styrofoam container filled with corn soup. Upon returning to his seat, he saw Ravi with fries dangling from the corner of his mouth.

'Man, yuh could ah buy some corn soup for meh!' said Ravi. He swallowed. 'Yuh know it is meh favourite.'

Surendra laughed. 'Here take this bag of popcorn.'

Ravi eagerly accepted the brown bag. He carefully placed it on his lap. 'Yuh could ah buy a bigger bag. Dis lil bag go finish soon.' He spotted a woman with a cooler. 'Marm wha' yuh selling dere?'

'Bake and buljol.'

'Doh make joke! Bring two for meh.' He smiled and began to dribble. 'Yuh know how long I eh eat dat!' He accepted the food and carefully balanced it on his lap. 'If yuh see anyone selling shark and bake tell dem to pass here.'

Surendra glanced at the back page of the *Trinidad Guardian*. The headline of the sports article was '90 minutes from Glory.' It would certainly be the longest ninety minutes in this nation's history. The newspaper pledged $11,280 to the football team for every goal they scored. Members of the public also pledged money for goals scored by the team.

Both teams returned to the field. The local team briefly found their rhythm but lacked constructive play. The creativity of previous matches was absent in this crucial game. The crowd was anxious. Superblue's calypso 'Dribble Dong Dey' was being played. Today it would be the main song in the country's impromptu celebrations. Fans with red and black painted on their faces looked despondent. They seemed like Caribbean versions of minstrels. The whistles and shouts of support continued but lacked lustre when compared to the confident voices in the first half.

The Trinidad and Tobago team could not break the solid defence of the United States. Dwight and Russell could not shake off their markers. Captain Clayton Morris needed to make a tactical change. The introduction of two substitutes- Hutson Charles and Maurice Alibey, offered a glimmer of hope to the fans. Hutson took a corner kick that went over the line and slammed into the side of the net. The net was calling for the ball. The Trinidad and Tobago team desperately needed a playmaker to create opportunities for a goal.

After ninety minutes, nervous faces glanced at their watches and the referee. A few more minutes of extra play had been added to the game. This was extra time due to stoppages during play. The drums begged for a goal. The clapping urged the local players to dominate ball possession. The whistles pleaded for a draw. The back of the net would never be found.

At 5.15pm the final whistle of the referee signalled the end of the short dream. The referee had become the hangman and released the trapdoor. Thousands were in shock. The equaliser never came from the Trinidad and Tobago team. Fans and players were crying. Some of the players were lying on the football field and either holding their heads or covering their faces. The end of the game meant the beginning of a nightmare- accepting the reality that Trinidad and Tobago was not going to Italy for the World Cup. It was a repeat of 1973 but on a grander scale. The grim dignitaries stared in disbelief. The vendors with their pockets overflowing with money were disappointed. Even the referee and linesmen seemed disappointed. The joyous World Cup fever had become a deadly plague. Only Ravi felt the day was a success.

Surendra looked at the lonely, disobedient ball on the field. Nobody cared about the ball which commanded the attention of tens of thousands of eyes for ninety minutes. Eleven heroes had been judged guilty. Eleven gods had failed their devotees who were in purgatory awaiting salvation. They could not get the one point needed to qualify. Eleven disciples who were worshipped by thousands had now failed miserably to cross the last hurdle in the long journey to Italy. They stumbled on the final step to paradise.

One goal had silenced the drums that were beating since the days of slavery and indentureship. One goal had taken the winds from the national flags. One goal silenced thousands of whistles. It was a day that tiny Trinidad and Tobago stood still. The local players made a burdensome lap of honour around the stadium's ground. Some of the heartbroken fans clapped and cheered. It was a charade but it was their elegy and consolation prize. Nobody offered sympathy.

Outside the stadium a few football fans were jumping and partying as if Trinidad and Tobago was victorious. Surendra looked at them in disbelief. He felt a sense of pity for these pathetic individuals who were celebrating amidst a grave loss. Only deranged persons would display this euphoria.

The next day was the national hangover. Persons drunk with enthusiasm and disappointment did not want to face the reality. It was a national holiday. A holiday in hell. Raphael dragged himself from the bed and went to the bathroom. He flushed the toilet and washed his face. He unlocked the front door and walked to the nearby parlour. The neighbourhood was still asleep. He purchased a newspaper and headed home.

He saw water near the refrigerator. He opened it and noticed some of the food was defrosting. He did not want to spend money on repairs. Two months ago one of the burners of the stove was not working. Yesterday the burner suddenly began working. He hoped the refrigerator would also repair itself.

The phone rang. He reluctantly answered it and heard crying. It was a woman's voice. He expected to hear Surendra or Ravi.

'Raf is Margaret....'

'Morning.' He did not feel like speaking to anyone.

'Is Ravi. He had a heart attack larse night and ah call de ambulance and dey carry he to de Port-of-Spain Hospital.' She loudly blew her nose.

Raphael was surprised. He momentarily forgot about the defeat in the football game. 'What ward?'

Margaret paused. 'Ward 29. It must be de game yesterday. He was too excited.'

'Okay don't worry. You geh some rest. Ah going this morning to visit him. Thanks.'

He replaced the phone on the receiver. He cursed and scratched his head. Raphael was in no mood to make a trip to the hospital. He looked at the headlines on the back of the *Trinidad Guardian* – "What Went Wrong." He was in no mood to read about the loss he witnessed. He should have stayed at home and view the tragedy on television. Yesterday he witnessed the crucifixion of an entire nation. The day for celebration was a day of mourning and regret. On Sunday 19 November 1989, God was not a Trinidadian.

Raphael turned the key and the blue Toyota Corolla came to life. The second-hand car was reliable and for the past five years did not need major repairs. The drizzle and overcast sky reflected the sombre mood of the country. After parking the car, Raphael purchased a lottery ticket from a nearby vendor.

'Give me a ticket with the number twenty-nine.' He placed the ticket in his pocket and headed for the hospital's entrance.

A female beggar approached Raphael. 'Boss help meh na. Ah eh eat yesterday and ah feeling to kill mehself.'

Raphael looked at the dirty and unkempt lady. 'Ent you does be by de traffic lights begging?'

'Ah used to be there. Buh de competition was too much. Odders came on meh turf and take away de customers. Dese days ah by de market.'

'Who yuh vote for in de larse election?'

It was an odd question but she decided to answer. She hoped that after all these questions she would receive money. 'De election officer didn't want me to vote even though ah had an identification card.'

Raphael looked surprised. 'Yuh name was on de voting list?'

'Ah cyah remember too well. He did arsk meh for proof of residence and ah return later with ah old cardboard box and a posey. Buh de police officers didn't want meh to enter de building with meh home. Boss ah real hungry, please help me na.'

Raphael did not want people to see him talking to the vagrant. He placed two dollars in her outstretched hand. 'Go and buy something to eat. No cigarettes or Bayrum.' He turned and proceeded to climb the stairs at the entrance.

She was elated. 'Ah going to buy food now.'

The woman headed for a food van selling noodles, fried rice and baked chicken.

As Raphael walked through the corridor, the smell of medicine invaded his nostrils. It was a sickly smell that he associated with hospitals. There were benches with old persons moaning. A pregnant woman sat on the floor. She shouted for assistance. Nobody responded to her pleas. It was a grim scene. Raphael shuddered and hoped he would never be ill in this hospital. He approached an obese nurse and asked for directions. The nurse had three huge bags filled with Panadol tablets. Raphael found the ward and entered.

Ravi was lying on his back staring at the ceiling. He was similar to the defeated footballers who were on the ground holding their heads. Ravi looked pale. He was thinking about the football game. The room had six patients. Two were sleeping and snoring loudly. One patient was calling for the nurse to bring a bedpan.

'Morning, morning.'

Ravi immediately recognized the familiar voice. He turned his head and forced a smile.

'Aye Raf how tings?' His mood was neither upbeat nor cheerful.

'Ah okay. Margaret call meh this morning and tell meh 'bout yuh heart attack.'

Ravi steupsed and became serious. 'She worry too damn much. Is jus a lil gas ah had. Some bitters and water or Andrews would ah clear it up.' He looked at the white plastic bag in Raphael's hand.

'Here ah bought yuh a card buh only I sign since everybody else was sleeping when ah left home.'

Ravi took the card and without reading it, he threw it on the ground. He shouted, 'Why de hell yuh bring Get Well card? Yuh could have bring some bags of chocolates. Yuh know ah like meh Charles and Cadbury.'

Raphael was shocked at his reaction. He picked up the card and slipped it into his pocket.

'Not even a blasted Kiss cupcake!' Ravi clutched the left side of his chest and began to breathe heavily. The exertion was too much for his body.

'Yuh want meh call de nurse or doctor?' asked Raphael. He had a worried expression and began to panic.

'No I okay.' Ravi stared at the ceiling. 'Is all dat stress of de football dat have meh so.' He paused. 'De Strike Squad is ah set ah big jackasses. Dem had de dreams of a nation…and jus so dey dash we dreams. De biggest jackass of dem all is de goalkeeper. Somebody should shoot him. All we needed was a draw to go to Italy. Jus' a draw. Was dat too much to ask?'

'What de doctors say was wrong with you?'

'Dem doctor and dem say ah have blocked veins...arteries or bones. Something blocked. Dey say dey go do some tests to determine if ah need overpass surgery.' Ravi paused and sighed. 'Or maybe is bypass surgery. Some surgery buh it sounding expensive. Dem doctors only like to cut up people and give dem wrong medicine. Thank God ah was never bright in school or else ah might have wanted to be a doctor.'

'Raf yuh have life and health insurance?'

'No. Rastas doh geh sick.'

Ravi was serious and stared at the ceiling. 'Doh play de arse. Life is unpredictable. Ah was once brave like you. Is only when ah had a long talk with Patty de doubles woman. Is she who advise me to geh life and health insurance. Thank God ah did lissen to she.'

'Patty? Where she selling?'

'Yuh muss know she. She does sell next to Ali's Doubles in Debe. Anytime ah go to see de madam family in Barrackpore, ah does buy eight doubles from she.' Ravi winced and licked his lips. He began to massage the left side of his chest. 'Man, ah feeling so hungry ah could eat six doubles. Yuh know dis country is ah strange place. Yuh doh need a food badge to sell food on de street. Look it does have people park on de highway selling watermelon, cascadura and roast corn.' He pointed to an empty bowl on a small table. 'Look breakfass dis morning was yellow jello. Yellow is not even meh favourite colour. Ah big man like me cyah survive on dat. Ah had to geh up from de bed and steal de breakfass from two other patients in de next room who still sleeping. And dem did get toast bread...and dey thinner dan me.'

'Dis is ah public hospital. Wha' yuh expect?'

He turned his head and looked directly at Raphael. 'Raf it not fair. Ah paying taxes and should be fed properly. Ah going to write de Ministry of Health and all de newspapers, ah going to call all de radio stations and complain.' He placed one hand on his forehead. 'Tings like dis does send meh blood pressure up.'

There was a soft knock at the door and Margaret entered the ward. She had a plastic cup and four brown bags. There were oil stains on the bags which contained pholourie, kurma, saheena and aloo pies. The cup contained peanut punch.

Ravi smiled. 'Yes, if yuh know how meh belly growling.'

Margaret laughed. 'Ah had ah feeling yuh was starving!' She smiled at Raphael. 'Thanks for coming.'

Raphael nodded. He peeped into the bin near to Ravi's bed. It had an empty carton of Supligen and wrappers of Sunshine Snacks, Zoomers, Chee Zees and Cheese Balls. It seemed Ravi had a death wish.

'Darling ah have to leave now.' She kissed Ravi on his forehead. 'Remember to lissen to de doctors and nurses.'

'Yeah, yeah.'

After Margaret departed, Ravi began to eat a saheena. He paused to lick the oily brown paper bag which contained the Indian delicacy.

When he was finished he licked his ten fingers. He then wiped his oily hands on the bedsheet. 'Yes, dis is how ah like it with slight pepper and plenty oil.' He then shoved his hand into a brown bag with kurma. 'Dese foods only taste good if dey oily or deep fried. De oil good for meh skin.'

Raphael looked surprised. 'Ent yuh have diabetes? Yuh should cut down on de carbohydrates.'

Ravi did not immediately answer because his mouth was filled with kurma. 'Yeah Raf doh worry after ah will take meh tablets for de sugar.' He pointed to a vial on the table. 'Is dat football game dat geh meh sick. Is after de goal score dat ah feel a pain in meh chest buh ah thought it was indigestion. Is only later in de night de pain come back again.' He inserted a straw into the top of a pack of Supligen. 'Ah wish ah had never gone to dat game. Latapy and all dem players is ah set ah haters. We should lock dem up in jail for stealing we hopes. Ah wish ah did never follow football, ah wish ah was a Grenadian or Barbadian and de coach should be fired a long time ago.'

Raphael interrupted him. 'Doh blame we boys and de coach. Dey try dey best. We should blame Meola de goalkeeper and the good defence of the United States. When ah tink 'bout it ah realize dat Indians not good in football. Look India with all dem people never never play in de World Cup.'

'Dat is not true. Football real popular in states like Kerala and Goa. And, Calcutta have two good teams, Mohun Bagan AC and East Bengal Club. In de Asian Games the Indian football team did win gold medals in 1951 and 1962. And at the 1956 Olympics de Indian football team was fourth.' He threw the empty pack of Supligen in the dustbin.

Raphael was surprised that Ravi knew these facts on Indian football.

'Ah should follow international competitions which Trinidad and Tobago not involved.'

'Yes, like de FIFA Club World Cup.'

'Yeah dat is good talk.' Ravi's eyes brightened and he became more upbeat. 'Meh fadder use to follow up dem games. And ah remember he had a folder with newspaper articles on dem FIFA Club games. Look ah remember in

1974 de club Atletico Madrid from Spain did beat Independiente of Argentina and in 1976 de club Bayern Munich from West Germany beat Cruzeiro which is from…Brazil.'

'Ent dey had other teams reach de finals like Feyenoord of Netherlands, Penarol of Uruguay, Benfica in Portugal, Panathinaikos from Greece and Malmo FF in Sweden.'

'Yuh well remember dem club names. Buh ah cyah remember what years and who dey play.' Ravi looked at Raphael's wristwatch. 'Raf dat is ah big wristwatch. Yuh ever thought 'bout getting a smaller watch. De damn ting looking like Big Ben clock in London.' He laughed then looked at Raphael's hair. 'Who give yuh dat haircut? It look like a blind barber use a dull knife to cut yuh hair.'

Raphael looked at the watch. He did not bother to respond. 'Yuh getting yuh kicks off me! Anyway yuh heard 'bout de gang violence in de secondary schools?'

Ravi was unconcerned with events in the country. 'No. Buh tell me wha' happen.'

'Students from El Dorado beating students from Hillview. And in Arima Senior Sec dey find students with guns and knives.'

Ravi replied, 'Football is one way to keep dem students occupied. Yuh know ah have to start watching more of secondary school football. It have real talent dere. Remember larse year dem three goals San Juan Secondary score in de final win in de Secondary Schools Football League?'

'Yes.'

'Buh meh favourite team in de North Zone is de Green Machine from St. Augustine Secondary. Dem is always ah favourite in Inter Col.'

The patients who were awake in the room were listening to the conversation. One was eating breakfast and the others were staring at the ceiling.

'All dat crime among students is normal for we country.' Ravi closed his eyes. 'Violence is part of us becoming a developed nation….we jus' have to accept it. Dem students from Hillview have to start carrying icepicks in dey bags to protect demselves. Lissen yuh see de patient to my left?'

Raphael turned his head. 'Yes.'

He whispered, 'Well he have dengue and de nurse giving him Panadol. And de odder patient near to him have tuberculosis and de nurse also giving him Panadol. De nurse also giving me Panadol but calling it by a fancy name- Paracetemol.'

There was a knock at the door. Ravi quickly hid the bags of food under his pillow. At the doorway was a slim lady with a dark complexion and elegantly dressed. Raphael thought it was a visitor or relative of one of the patients in the room. Ravi was relieved.

'Eh, eh honey ah came to surprise yuh!' said Renata.

Ravi wiped his mouth. 'Renata, come leh meh give yuh a kiss.' He introduced her to Raphael. 'This is my good friend Raphael who came to visit me.'

Renata shook Raphael's hand. Raphael watched her closely. She spoke with an accent and was of Mixed descent.

'Here honey, ah brought a chicken roti for yuh. It have slight pepper.' She removed it from the bag and carefully placed the roti on his belly.

Ravi gently patted the roti. 'Thanks darling. Ah starving here. Is like if ah in jail. De feeding meh jello…yellow jello and red jello. Imagine prisoners does eat better dan dat.' He then lifted the roti and said, 'It feeling nice and heavy ah hope dey give meh some nice pieces of meat and not neck.' He kissed the roti and returned it to the spot on his belly. 'God bless you.'

Renata sat beside him and began to comb his hair. She then wiped his face with a small towel. She pinched his cheeks and fondly touched his chin. 'Honey yuh looking so pale. When yuh come out here ah want to cook some macaroni pie and stew chicken for you.'

'Darling yuh could ah bring de macaroni pie and chicken today. Wha' de hell ah go eat for dinner today?'

Renata laughed. 'Ah have to leave now. Ah have ah meeting next hour.'

After Renata departed, Ravi sat upright on the bed. He began to eat the roti. He looked at Raphael who was patiently waiting for an explanation. 'Ah know yuh must be want to know who she is. She is a very good friend.' Ravi sighed and stared at the ceiling.

'Wha' yuh mean by good friend?'

'De outside woman. Ah deputy nah.'

Raphael was disappointed to hear those words. 'How yuh could do dat to Margaret? She does real care for yuh.' He moved some hair that had fallen on his forehead.

'Margaret know 'bout she. Margaret know dat ah go never leave she and go with Renata.'

'Yuh have to be loyal and faithful to one person. Stop horning Margaret.'

Ravi stopped chewing and pointed to Raphael. 'Doh give meh no damn lecture. Yuh living common-law with Juanita. Suppose ah start accusing you of living in sin? Me and Renata have a special relationship where she does cook for me in exchange for a lil something.' He opened a box of Sunday Basket and poured ketchup on the chicken and fries. He licked his lips.

Raphael remained quiet then shook his head. 'Anyway is your life. You is ah big man. Ah have to go now, ah have to buy a few items in de drugstore.'

Next day Ravi underwent the procedure of triple bypass heart surgery. Raphael and Margaret visited him in the evening. Ravi was unable to respond to anyone. He had a ventilator to assist breathing and he was heavily sedated. During the next six days he remained in the Intensive Care Unit. There was a scar on his leg and middle of his chest.

Raphael decided to visit his aunt, Barbara, who was also a patient in the hospital. She spent most of the last fifteen years in Ward 6. Doctors diagnosed her as a hypochondriac because she regularly complained of imaginary pains and symptoms. Her insurance company regretted having her as a client. Raphael felt Barbara liked the attention of nurses and doctors in the hospital. He felt she should have been at home since the hospital never had sufficient beds.

Barbara was diabetic and also suffered from high blood pressure. 'Raf come and kiss me, long time yuh didn't come and see me.'

Raphael reluctantly kissed her cheek. He was concerned that there were germs on her face. Her hair was neatly plaited and she smelled of Limacol. There was a thick layer of Vicks and white powder on her neck.

She pointed to her bandaged right foot. 'Lissen yesterday de doctors and dem cut off meh big toe. Dey say is de sugar cause it. I did step on a nail and piece of glass and didn't even know. It stay in meh toe for a few days. And, yuh know it doh have any good doctors in Crown Trace.'

Raphael raised his eyebrows. 'Ah sorry to hear dat. Untreated diabetes is a dangerous ting. People does geh blind and go in coma and ting.' He wanted an excuse to leave the room and hoped a nurse or doctor would soon enter the doorway.

'Lissen yuh coming to de funeral?'

'Funeral? Who dead?' Emmanuel asked. He thought it was a recent death. 'Ah didn't hear anything or see it in the death notices.'

'Is for meh big toe. Tonight and tomorrow there will be a small wake service under meh house and Thursday morning is de funeral. George renting a tent for de wake, he order plenty food and already pay de mike man to announce it today in de village.'

Raphael wanted to laugh aloud. He gently bit his tongue and had a poker face. 'Ah will certainly try and come. Burial or cremation?'

'It will be buried at Paradise Cemetery in a plot near to where meh little toe and left leg buried. Ah feel ah will dead soon, since two years ago ah have meh funeral clothes airing out. Ah already pick ah burial spot and done choose ah tombstone.'

'Tantie yuh have long to live, doh worry.' Raphael knew she was always planning her funeral. He joked to relatives that Barbara was a professional patient.

Barbara was serious and closed her eyes. 'Ah told George to call all meh chirren in Canada, America and England and tell dem to come dong for de funeral on Thursday. If ah survive den dis will be my larse Chrissmass.' She patted her chest and looked glum.

'Yuh hear dat stupid gyul Belinda geh bounce down yesterday evening while crossing de highway and she dead. De family living near de highway and just larse year she fadder geh bounce dong too on de same highway. And it did happen to she sister also. It look like cobo pee on dat family.'

'What! No, ah didn't hear dat. Ah know de whole family recently start wearing fluorescent clothes so drivers would see dem,' said Barbara. 'Belinda was de soloist at de funeral of meh lil toe. Buh how de family go afford de funeral? Dey poor buh dey should sue de driver.' She passed her hand over the creases on the cotton blanket.

'Well de driver of de car own a funeral home so instead of legal action, de family will accept his offer of a free casket.'

Barbara glanced at the clock on the wall and nodded. 'Dat was a good choice. Ah would ah take de casket too. Ah hope it is ah expensive, fancy one. Give my condolences to de family. Anyway nice of you to visit, ah have to take meh rest now.' She closed her eyes and gently placed both hands on her chest.

Raphael bid her farewell and left the hospital.

One week after Ravi's surgery he was returned to the regular ward and visitors were allowed to see him. Raphael was his first visitor.

'Yuh could have brought some chocolates and beer. Ah feeling extra hungry. Yesterday ah vomit and….' Ravi began to cry. He had lost fifty pounds during his stay at the hospital.

Raphael interrupted him. 'Boy, yuh jus' had a close call. We almost lose yuh…yuh have to start eating healthy.' He cleared his throat.

'De damn doctors and dem take a vein or something from meh leg and puh it in meh chest.' Ravi paused and wiped his eyes. He pointed to his right leg. 'Look at de ugly mark on meh leg and it feeling weak. Ah doh tink ah will be ever able to play small goal football again. And ah will never be able to swim properly.' He continued crying.

'Doh worry, yuh could be a defender or goalkeeper since dat hardly have much running in small goal. You could swim? Where yuh learn to swim?'

Ravi stopped crying. He drank some water then carefully placed the glass on the table near the bed. 'When ah was a small boy living in San Juan, every year use to flood and is right dere in de flood water ah learn to swim.'

'Buh dem flood waters dangerous. People does drown in it.'

Ravi smiled. 'Nah in San Juan it had clean and safe flood waters. It wasn't juss me. De neighbour chirren, goats and ducks was also in de flood waters. After de floods we always geh food hampers and a new mattress from de government. And each farmer use to geh only five dollars to compensate for dere crop losses. Thank God for the Ministry of Mattresses and Hampers. Dis prime minister will never solve flooding but ah hear he planning to offer swimming scholarships and free swimming lessons to people who live in areas dat flood.'

'Hear na so how de shoes selling?' asked Raphael.

'Dat is one of the worse business deals ah ever make.' Ravi had imported five hundred pairs of high-heeled shoes from China. All were the same colour and size. After four years he had sold three pairs. He decided to give shoes as birthday and Christmas gifts but them shoes were never used as women found them too uncomfortable. 'When ah come out of here, ah will throw away all. Is true advice yuh did give meh 'bout checking de market before getting into business like dat.' He cleared his throat. 'Buh Raf wha' dey do with de clogged vein from meh chest?'

Raphael was curious. 'Ah believe it is ah artery. Why yuh want to know?'

'Is MY vein…and ah want it. Ah did ask de doctor after de operation buh he was from Cuba and could only speak Spanish and de nurses from de Philippines. Three years ago dey take out meh appendix and ah never geh it back.' Ravi had a tone of desperation in his voice. 'If ah doh geh meh vein ah plan to carry de hospital to court. Raf if somebody geh meh vein dey will do obeah on me. Yuh will help meh geh meh vein?'

Raphael shook his head and stared at the floor. 'Yuh better ask somebody else, ah busy dese days. Ah have to buy clothes to attend a wake service and funeral dis week.'

Chapter 3
City and church

Raphael's wife, Juanita, was of Spanish and African descent. She owned a flower shop in Montrose Mall. Juanita was twenty years old and had a light-skinned complexion. She was described by Raphael's parents as being 'high colour.' The couple was blessed with four children- Marcus, Malcolm, Cleopatra and Harriet. They were named after important Black persoanlities- Marcus Garvey of Jamaica, Queen Cleopatra of Egypt and Malcolm X and Harriet Tubman of the United States.

'Ah fed up tell yuh dat one day yuh will go in jail for a long time for smoking ganja!' She slammed a bucket on the kitchen table.

Raphael calmly responded, 'Doh worry 'bout me. Dis is de holy herb. Jah go protect me.' He took some of his dreadlocks that were hanging near his face and flung them over his head. He looked at the ceiling and slowly released smoke from his nostrils. He murmured, 'Who Jah bless leh no man curse.'

'You stay dere and feel so. Ah doh want to give yuh de length ah mih tongue. When dem corrupt police ready for yuh, even if yuh doh have ganja on yuh dey go plant some on yuh. And everybody in court go believe de policemen when dey say dat de Rasta man selling and smoking ganja. And ah fed up tell yuh dat we and de chirren go geh lung cancer with dat smoke. Go outside and smoke.'

Raphael steupsed loudly. 'Is cigarette smoke dat bad. Dat have man-made poison.' He shoved the joint in her face. 'Dis is natural. Dis is de holy herb. People does use dis to cure asthma, cancer and reduce high blood pressure.'

Juanita shouted, 'Save dat for de judge. Doh forget to call someone to fix de fridge. It leaking and not keeping de food cold and the kitchen sink backing up. Wha' is dat yuh writing?' She took a cocoyea broom and began sweeping.

'Doh worry it was giving trouble larse year and it fix itself. Ah jus' writing a lil eulogy for Uncle George to read.'

She steupsed. 'Yes dat is wot does have to happen in dis house- appliances does fix demselves. Ah fed up tell all yuh not to wash dong rice grains and chicken bones when yuh washing wares. Dat cause de pipes to clog up. Remember meh mudder and sisters coming to visit this Sunday and next week is meh two cousins.'

There was always an argument whenever Juanita raised the topic of visiting relatives. Twice a month there would be visits from her relatives.

Raphael muttered, 'Every weekend is some damn relative visiting. Is best ah did choose ah wife from de orphanage without any parents and relatives.'

She heard him. It was a statement that Raphael would soon regret.

'Who de hell yuh tink yuh speaking 'bout? Yuh feel meh relatives is some ole niggas who does smoke weed?'

He remained quiet. He knew the argument would be prolonged.

She began to slam the wares in the sink. 'Every week one of yuh good fuh nutting friends or familee does land up here with dey empty belly and expect me to feed dem.'

Raphael left the shack and headed for the hammock which hung between two mango trees.

Whilst in the hammock he looked at the exterior of the shack. It had four rooms and he planned to build another room nearby to serve as a garage. The walls of the shack comprised thin sheets of wood that he obtained from a nearby sawmill. Fifteen rusty galvanize sheets served as a roof and these needed a coat of paint. On the eastern side of the roof was a mossy piece of plastic spouting. At the corner of the house there were two iron barrels which were filled with rain water from the spouting. The water in these barrels would be used for cleaning dishes, watering plants and showering. The home had electricity as a result of an illegal electrical connection.

After twenty minutes he decided to check his plots with cassava, corn and pumpkin. The crops needed watering. He took the empty Milo can and filled it with water from one of the barrels. Raphael thought about the recent argument with Juanita. He did not like to argue and knew that she meant well. They had been in a common-law relationship for five years. During those years he had neither hit her nor committed adultery. Juanita's parents always encouraged him to get married but he felt happy in the relationship. Juanita wanted to get married because she felt the union would be stable and their children would be legitimate. Raphael justified their unwed status by claiming that most married men and women often disregarded their sacred vows. The couple was unmarried but annually celebrated 19 July as the anniversary of the day they began their common-law relationship.

Juanita emerged from the shack with a plastic basket filled with wet clothes. She looked at him watering the plants. She was angry that he wanted to remain a squatter. He did not seem to have any future plans of buying land and building a permanent home. She stood at the stand pipe and washed the clothes with a bar of blue soap. After, she rinsed the jookin' board and headed home. Between the plum and sapodilla trees were two clothes lines. She hung the wet clothes on the lines and hoped the day would be sunny. His recently washed jerseys and pants were hanging on these lines. The scrubbing board was left outside to dry and she returned to the shack.

Raphael walked to the back of the shack to see if Marcus was cleaning his crepesoles with a rag.

'Dread rub dem shoes hard till all de mud come out. Yuh cyah be going out with black shoes dat covered in mud.' Raphael threw away the small joint of ganja.

Marcus rubbed harder. 'Yes daddy.' He was an obedient boy.

Raphael looked at his young son. He wanted to give him advice. 'Make sure and bathe every day. You and soap and shampoo must be friends. Doh lissen to what people say dat Rastas doh bathe. Keep yuh dreadlocks clean and have it tied up neatly when yuh in public. Puh it in one of de caps I buy for yuh. Yuh understand? And, doh smoke ganja like me…yuh will geh in trouble with de police. Become better dan me.'

'Yes daddy. Why yuh telling me all dis?'

'Yuh eh no lil boy again, ah not going to speech off you and yuh sister again. Yuh is ah big man. Ah want allyuh to be respectable citizens in society. Doh ever feel inferior.' Raphael knew that his children were mocked at school because he was a Rastafarian. He sat on a stool and began shaving with a razor blade. His face did not have any lather.

'Yes daddy.' He stared at his father. This was not simply advice from his hero and mentor. It was an important moment of bonding between father and son.

'Chief, tell me how school going. Everyting under control?'

'It okay.' Marcus paused. He thought about an answer that would not create an argument. 'Mathematics a little hard and ah doh understand de Physics.'

'Why it hard and yuh doh understand? You have to mash it up! Yuh have to fight up. You is de nex' generation.'

'Well de Mathematics teacher does only read from de textbook and Eddie Fung Woo de Physics teacher hardly come to class because he umpiring cricket matches or attending union meeting or staff meeting.'

Raphael laughed. 'And for Parents' Day, de Physics teacher boasting to everybody how he is ah professional. Doh worry from nex' September ah will find a lessons teacher for you in dem two subjects. Dem teachers is ah waste of time all dey doing is collecting a salary and frustrating chirren. When ah was small like you de teachers use to waste time and ah was shame to tell people ah still in school…so ah told dem ah was unemployed.' He stopped shaving and passed his hands on both cheeks. 'Meh face hard like leather. When ah die use meh face to make some shoes or a belt.'

Marcus smiled. He knew his father was joking.

Raphael returned to the hammock and lit another joint of ganja. He watched the blackbirds pecking at half-ripe Julie and Starch mangoes. He got up and placed coals under the coal pot. He added rice, chicken and vegetables to the pot. Tomorrow he would climb the trees and pick mangoes. Some would be sold in his stall. Juanita preferred the half-ripe mangoes. She would make chow by cutting them into small pieces and adding salt and pepper.

The screech of brakes and blaring of a horn startled Raphael. Loud rap music broke the serenity of the afternoon. Juanita had a cobweb broom and was standing on a makeshift ladder. She moved the curtain and peeped outside. She thought it was the police.

Raphael continued smoking and ignored the commotion.

Surendra shouted, 'Raf! Raf! Come with we to tong!'

He removed the ganja from his mouth and slowly turned his head. He saw his friends and smiled. 'Why all yuh doh drive like normal people?'

'Raf go put on a jersey and jump in de jitney quick. We going tong!' shouted Ravi.

'Jus' so? Wha' tong have dat better than h….'

He was interrupted by Surendra who pointed in a northwesterly direction. 'We going to loot. De police headquarters bun dong and gunmen takeover Parliament.'

'Right ah coming now. Leh me geh a jersey and tell Juanita.' Raphael ran towards the shack. His waist-length dreadlocks swayed from side to side. As he opened the door he met Juanita who appeared concerned.

'Wha' is de excitement out dere?'

'Ah going to tong with de boys. Dey say it have looting of stores and ting and we going to see what we could get.'

'Good. Bring a jeans, size thirty three. A pair of shoes for Marcus and school bags. Doh worry about toys. Remember school opening in September.'

Raphael began to search through a pile of clothes on the couch. He found a yellow jersey and slung it across his right shoulder.

Juanita stopped him. 'Raf how yuh could wear dat ole jersey? Yuh going out in de public so wear something nice.' She opened a drawer and selected a red and yellow jersey. 'Here take dis it have a picture of Che.'

Raphael nodded. He accepted the jersey and returned his choice to the pile.

'Raf remember if you see any jewelry bring it na. Next month is we anniversary.'

'Yeah ah cyah forgeh dat.' He closed the door and jumped unto the jitney's tray.

Surendra slipped another cassette into his tapedeck. The music of Bunji Garlin flooded the air.

'Aye turn dat ting up! After play some rockers,' shouted Raphael.

At Aranguez, they saw clouds of thick black smoke rising from Port-of-Spain. There was a long line of traffic on the opposite lane of the highway. Anxious drivers blew their horns. Ravi wanted to urinate and decided stop on the shoulder of the highway. He left the vehicle and headed for the bushes.

Raphael laughed. 'Dis country is de greatest. Anybody could just stop de car and pee on de highway. Yuh go never see dat in North America or Europe. De police should pass and charge yuh for littering.'

At the traffic lights, in Aranguez, there was a beggar with a deformed left arm. He had a crocus bag slung over his shoulder. It contained empty bottles which would be sold at the glass factory for a few dollars. He was homeless and would spend his nights in Tamarind Square. During the day he would kick pebbles and remember his childhood days when he played football and was contented. For fifteen years he stood at the traffic lights with an outstretched hand and walked near to each car. Most motorists ignored him. The traffic lights turned green. Ravi changed gears and sped towards the city.

Raphael was annoyed. 'Aye slow down, we not in a hurry. You have lead foot?' At the next set of traffic lights, there was a beggar with a limp. During the past decade his home was the carpark at Riverside Plaza. This was an area inhabited by the city's homeless and poor. He had a club foot and three fingers on his right hand were missing. His wrinkled forehead and distorted face reflected many years of agony. His graying hair was sparse and uncombed and his long beard was dirty and untrimmed. There were holes in his buttonless shirt and pants. Nearby was a police vehicle parked on the shoulder of the highway. Two policemen were urinating in the bushes.

The beggar slowly approached the van. 'Wha' yuh have for yuh boy?'

Raphael asked him, 'Ent you use to be by de next traffic lights?'

'Yeah, buh meh brudder helping out dere. So ah decide to come by dis one.' His foul body odour indicated he did not have a decent bath for many months.

Ravi smiled. 'So you expanding de business. Yuh open up ah branch over here?'

'Sort of.' The beggar forced a smile. He did not detect the sarcasm and was accustomed to insults and sarcasm. He coughed. 'Yuh have any change?'

'Have you asked the government for help?' asked Ravi.

'Yes dey does give meh a cheque to help buh ah need ah lil change to buy lottery and odder pussonal items.' He scratched his head.

'Well we doh have any spare coins buh go to de city. See dat smoke over dere. Come and see if yuh could geh something. Ah hear things giving away,' shouted Raphael. 'We going dere now.'

The lights turned green and the van left the beggar at the traffic lights. The beggar watched the van depart for Port-of-Spain. He turned and saw huge clouds of smoke emanating from the city. He looked at the cars speeding past him. He often had suicidal thoughts and wanted to end his miserable life. He began walking to the city.

The van with Raphael and his friends finally reached the city. Port-of-Spain was chaotic. Some frightened citizens were shoving to get buses and taxis whilst others were rushing into stores to steal items.

Ravi parked the van near the Port-of-Spain bus terminal. 'Is better to park here so when we ready to leave it go be easy. All yuh take half an hour and let's meet right here.'

'Dis door cannot open,' said Surendra.

'Yeah, dat giving trouble a long time now. Roll dong de glass and pass through the window.'

Surendra followed Ravi's instructions. Surendra was slim and this was not a difficult task. It was an odd method to exit the vehicle but nobody in Port-of-Spain noticed or cared.

Raphael hurried to Frederick and St. Vincent Streets. There was fire in most of the buildings and the sounds of glass being broken filled the air. Frightened people wearing businesses suits were running from the city. Hundreds of persons were looting stores. Capitalism had gone mad.

A young man had a large television on top his head. He was struggling to move with the heavy load. Raphael entered Francis Fashions and began searching for jeans and children's shoes. He grabbed pairs of Air Jordan, Puma, Nike and Adidas shoes. He saw plastic bags and boxes on the counter. The looters had no need for bags. Many were quickly snatching items and leaving the stores. Raphael took five plastic bags and quickly filled these with clothes and shoes.

One looter set fire to the entrance of an appliance store. Raphael shouted at him, 'You really dotish. If you set fire to de entrance how de hell yuh expect people to get stuff? Yuh mean somebody have to teach yuh de proper way to loot?'

Looting appealed to low income and unemployed persons. For sixteen years, Thomas, a vagrant, begged outside stores. He was dressed in torn, smelly rags. The owners constantly chased him away from the entrances of their stores. Today, Thomas would be king of Port-of-Spain. He saw smoke but did not know about the bombing of the police headquarters and the attack on Parliament. All he knew was that today he had the freedom to enter clothing stores, fast food outlets and groceries. For many years he slept on the pavement under various stores. Many nights he would dream of entering these stores. Today these dreams would be fulfilled. He did not know about the laws of supply and demand. He never had a bank account and never cared about sales and discounts. Today he fully understood that everything was free.

On hot days he would walk in the shallow drains to cool the blisters on his feet. Today he would listen to owners begging for protection from the police and army. Today he had a smile on his face. Today he would not have to dig in dustbins for half-eaten and decaying foods. Today, Thomas would not have to search for discarded drinks and food that had maggots, ants and flies. He could now enter a Chinese restaurant or Royal Castle and obtain fresh food. His city that scorned and mocked his existence would be scorched.

Surendra looked at the overflowing dustbins which were bolted to the pavement. He was surprised that this project was undertaken by the government. He wondered who would steal dustbins. He wondered why the government was so concerned with protecting the dustbins but ignored the security of the country.

A maxi-taxi with looters and their stolen goods departed Port-of-Spain. Taxi drivers assisted looters in loading stolen items into backseats and trunks. A man attached a chain to a van and the other end was wrapped around security iron bars of American Stores. He drove forward and the bars were ripped away. Excited, screaming looters rushed into the store.

There was a loud explosion near Woodford Square. An engineer who was running through the Square was startled and dived into the nearby water fountain. Unfortunately there was no water in the fountain and he dislocated his right shoulder.

Persons in respectable jobs including bank clerks, secretaries, teachers and receptionists were also in the melee. They were attracted to the availability of free items. They were running from store to store and grabbing as many items as possible. Today's attack on democracy would be a concern for tomorrow.

Port-of-Spain was noisy. The air was filled with shouting and blaring security alarms. The smoky air did not seem to bother the looters. Surendra noticed covers were missing from some manholes on each pavement. He wondered who would steal the metal covers. He was amazed that looters and hundreds of panicked citizens did not fall into these open manholes. During the nights, Thomas would remove the city's manhole covers and sell them to a scrap yard dealer in Belmont.

Raphael wished he was still at home. He suddenly missed the quiet afternoons under the mango tree. Surendra entered a store and departed with a microwave, video recorder and radio. Ravi had a washing machine on his back and slowly walked to the bus terminal. Inside the washing machine were two footballs, a pair of football boots and four shin pads. He had trouble breathing and would occasionally stop to massage the left side of his chest. In his pockets were cassettes of Buju Banton and Sizzla. He gently placed the washing machine on the van's tray and covered it with green tarpaulin. He sat next to it and waited for the others.

A short, plump woman was running with three rolls of coloured cloth under her arm. She shouted to everyone, 'Dis is to make curtains and bed-sheets.' She stopped shouting when she saw heavily armed soldiers from the Army. A policeman entered a pharmacy and opened the cash register. He filled his pockets with twenty and hundred dollar bills.

Raphael shouted at the policeman, 'Babylon yuh always feel yuh is ah gorgon and high ranking. Always downpressin de brethren. Dis dread sight up. Seen!'

The policeman heard the remark. He felt guilty and quickly exited the store and headed for a taxi.

Raphael muttered to himself, 'All dem kinda people could do is tief, make chirren and waste money.' He entered a store with appliances. He saw his cousin, Ariana, from Belmont. She and her two sons came to loot in Port-of-Spain. One son had two rugs and a water pump whilst the other son had a tuxedo and camera.

'Blessings,' he said. He placed his bags on the ground and embraced Ariana. They had last seen each other at her husband's funeral in 1988.

Ariana was elated, 'Long time we eh meet. Yuh know, dis is de furse time dis year dat de family making a trip to de city. We were so busy for the past few months buh two days ago meh mind run on you. Yuh go live long. How Juanita?'

'The queen is fine. She home with de chirren. Tell Oscar ah say hello and tell him to change he life and accept Haile Selassie as de messiah.'

'Yuh know Oscar is ah rude bwoy. He only want to pardy.' She began to shake her hips.

Raphael uttered a loud laugh. He never liked Oscar, her common-law partner, because he was arrogant and boastful. 'He is ah bald-head rasta. He like lazy man wuk and always like to do name-dropping, telling me he know dis gangsta and dat politician. Tell him he wasting he life with dat pardying and friends who want to paardah. He have to smoke the weed of wisdom or else he go end up in a wooden suit.' Raphael paused. 'Yuh know ah jestering.'

Ariana laughed and poked his stomach. 'Yuh is always a joker.' She saw the bags in his hands. 'So yuh geh plenty nice tings?'

'Too much.' He looked at his watch. 'Lissen we go talk later, ah doh want to miss meh ride. Guidance.'

'Right. Bye Raphael.'

Juanita was in the kitchen peeling eddoes and cassava. She was making a soup. She had already peeled four yams. She was worried that Raphael had not returned home. She tried humming a hymn. This was interrupted by questions from her children.

'Mommy where daddy gone?' asked Malcolm.

'When daddy coming home?' asked Marcus. 'He have to help meh with a puzzle.'

'He gone shopping in de city to geh nice tings for we. All yuh watch Twelve and Under and not to worry.' She stood by the sink and washed the vegetables. She cut the vegetables and placed them in an enamel pot on the stove. She opened the refrigerator. There were no carrots. She always prepared foods that were fresh and natural. Raphael ate I-tal food which did not contain preservatives and colouring.

Cleopatra shouted, 'Mommy come de news start. Ah man with a cap reading it.'

Juanita glanced at the clock on the wall. It was not 7pm. She felt it was odd that there would be a news reporter wearing a cap. She switched off the stove and went into the living room.

On the television screen were men in Islamic garb and seated at the side of two news announcers- Dominic Kalipersad and Jones Madeira.

She did not recognize the Muslim men. 'Turn de volume up.'

Malcolm quickly turned the knob and increased the volume. He also had a sudden interest in the news.

'The government of Trinidad and Tobago was overthrown,' said a bearded man. He explained Parliament was not functioning and that a new prime minister would soon be appointed. The man appealed for calm.

'Wait. De man at the back of Dominic is Marcus uncle! Dat is Jameel!' After the short broadcast Juanita dialled two friends to relay the distressing news. She returned to the kitchen and struck a match to light the old, gas stove. Her hands were trembling. She turned the dial of the radio and listened to Radio 610 AM. She was very worried that her partner had not returned home.

At eight o'clock Raphael returned home. His pockets were bulging and his hands strained with overflowing plastic bags. He looked like a Black Santa Claus with gifts. Cleopatra, Harriet, Marcus and Malcolm rushed to greet their father. Juanita had a shocked expression on her face. She was relieved that he was unhurt. She hugged him and began to cry but he brushed her away. He wanted to display the loot.

'Ah want all yuh to see dis.' Raphael emptied the contents of the bags. There were socks, copy books, jeans, shoes, caps, blouses, jerseys, a radio and cutlery. He dipped into his pockets and pulled out gold bracelets, diamond pendants, gold rings, pearl earrings, wristwatches and gold chains.

The four children were excited. Tonight their father was Santa Claus and a magician. The family would never forget that night.

The soup on the stove was bubbling.

'Honey wat yuh cooking? Soup?'

Juanita was in a daze. 'Yeah. It almost ready. And if yuh want ah could fry some plantains.'

Raphael sat in the dining room. He was exhausted. 'Oh Jah! Ah never wuk so hard. Honey ah hungry. Switch off de stove and bring some of dat soup for meh. Today ah wuk hard.' He softly said to himself, 'Ah wish daddy was here to share in all dis ah bring home.'

Cleopatra and Harriet entered the dining room. They wore over-sized jerseys and shoes. Cleopatra wore gold chains and Harriet had bracelets. Marcus was attempting to tie the laces of his new sneakers. He was now eager for school to re-open in September.

Malcolm had two caps on his head. He read aloud the logos on each jersey, 'Welcome to New York', 'I'm a rude boy,' 'Bob Marley the legend'. He was excited to see his name on one jersey which also had a large photo of Malcolm X.

Juanita filled a bowl with soup and entered the dining room. She carefully placed it on the table. 'Be careful it hot. Your Aunt Barbara called this evening. She said de funeral was big and dat everyone like de eulogy dat Uncle George give. Buh yuh didn't tell me dat one of yuh family had died.' She returned to the kitchen and began washing the dishes.

Raphael placed a mouthful of soup in his mouth. 'Doh worry is somebody small. You wouldn't know him.' After two mouthfuls of soup he decided to let it cool before continuing. He shouted to Juanita. 'Lissen now dat we a lil richer ah want to marry yuh in ah church...make it official.'

Juanita dropped a plate she was washing in the kitchen sink. She quickly entered the dining room. 'Yes dear, and we can use the rings yuh geh today as de wedding rings.'

Next day Raphael went to buy a newspaper. The vendor indicated that she did not receive any copies that morning. Raphael returned home and switched on the television. There was grainy footage of the looting in Port-of-Spain. He saw a lady dragging a blue stove through the streets. Raphael recognized her. This was the beggar who approached him at the hospital. She had forgotten her name and was abandoned by relatives. For many years the city's pedestrians and motorists avoided her. Today she was part of history and did not care to be remembered.

Five young boys were running with couches, a microwave oven, a mahogany table and two teak chairs. Dishonest soldiers were also looting. Six soldiers entered Standard distributors and seized blenders, lamps, fans, mirrors and beds. Soliders entered Tru-Valu and stole frozen turkeys, tinned foods and expensive wines.

Raphael saw a middle-aged man dragging a refrigerator. The man's hair was uncombed and his fingernails were long and black. He wore an expensive Rayban shades, new jeans, a gold chain and his long-sleeved Van Heusen shirt was unbuttoned. Raphael did not know the man whose name was Mark. During their trip into Port-of-Spain, Raphael and his friends had briefly met him at the traffic lights.

Mark was the beggar with a deformed hand who spent most of his life begging at the traffic lights in Aranguez and Curepe. Today he did not have his crocus bag. For fifteen years nobody had referred to him by his proper name. When he was younger, some persons laughed at his epileptic fits. The world had forgotten Mark and he had ignored the world's achievements. He had been exiled and banished from Port-of-Spain but in 1990 returned to judge the uncaring city. He never thought about the concepts of democracy and freedom. He neither possessed a house nor electricity. He planned to place the refrigerator near a dumpster and use it to store his tattered rags. These rags offered little protection during the cold, rainy nights.

Some Trinidadians were worried that the coup would result in food shortages and power outages. They flocked to the groceries and bought batteries, torchlights, candles, frozen meats, tinned peas, milk, huge bags of flour, rum, soft drinks, rice and potatoes. Kitchens had overstocked shelves and garages resembled warehouses. There were long lines at the gas stations. Drivers feared gas shortages and wanted to ensure their tanks were filled.

Raphael switched on the radio. He heard part of an interview with Harrilal, the leader of the coup.

Harrilal arrogantly proclaimed, 'There is no unrest. There is no looting. Ask all those people calling my name and clamouring for me in the streets if this is bad! Who gave them food, money and medicine? Not the government. I am a loyal citizen of Trinidad and Tobago. I am the leader of a just revolution. There must be no looting. What we want is peace in the land. This is a popular revolution of citizens of dis country!'

Raphael laughed. He was now convinced that Harrilal was a lunatic. Many young Trinidadians neither cared nor understood the seriousness of the coup, the trauma endured by the hostages and the dangerous threat to democracy. The government had enforced a curfew to restore order. This restricted the movement of citizens. Ravi, Raphael and Surendra organized and attended curfew parties. Friends would arrive during the evening, before the start of the curfew and leave the party after the curfew ended at 6 am. Nothing could prevent Trinidadians from partying and enjoying themselves. A few disobeyed the curfew and had to serve a short jail sentence. These curfew breakers were placed in cells with rapists and murderers. The coup soon ended. It was another seven day wonder of the island.

In October 1990, Ravi began a daily exercise routine of walking to reduce his weight and maintain a normal sugar level. This exercise programme was suggested by the doctor to ensure Ravi's diabetic condition would not worsen. He would walk from 6am to 7am and then proceeded to his job. After work he would walk for one hour in the evening beginning at 4pm. During the weekends he would walk from 1pm to 4pm. It was a rigourous schedule.

One evening, Raphael decided to walk with Ravi. It was overcast and the clouds indicated the possibility of rain.

'So who yuh tink was the hero of this World Cup?' asked Ravi.

'This was a game where dere is always more than one hero.' Raphael quickened his pace. 'I like to see Ronald de Boer de midfielder of Holland play. He was a live wire in de team dat won de European Cup, Dutch Cup and UEFA Cup.'

'Ent he had a brother or relative who also played for Holland?' asked Ravi.

'Yeah buh ah cyah remember his name. Dis game had good defenders like Iharu of Japan and Ivanov of Bulgaria. Scotland has a left-sided midfielder. I tink his name was Collins and he was good with freekicks. Larse year he scored a beauty but the linesman disallowed it.'

Ravi shook his head. 'Yes John Collins. He was also good at delivering long crosses to strikers. Ah expected more goals from strikers such as Ali Daei of Iran, Solskjaer from Norway and Molner of Denmark.'

'You know who was de most entertaining?' asked Raphael. He looked at his watch. It was ten minutes past three o'clock.

Ravi laughed. 'Yes, is obvious- Milla of Cameroon! People who saw dis World Cup will not forget his dances around the corner flags after he score his goals.'

Raphael agreed. He looked at the piles of black garbage bags. These were placed near his hand-painted sign – No Dumping of Rubbish. He wondered why Trinidadians never obeyed laws and signs. 'Lissen, next week ah going Tobago with some friends to see Goat and Crab racing and while ah dere will do some windsurfing. Why yuh doh come? We staying at the Coco Reef Hotel and two of meh friends will be at Grafton Resort and plan to participate in de Presidents Cup Golf Tournament which will be at Tobago Plantations.'

Ravi had a smirk on his face. 'To tell yuh de truth ah fed up go Tobago, remember ah does be dere almost every year for de Great Race. Ah have a cousin who working at de Trinbago Curry House and Bar at Pigeon Point Junction and meh uncle own de Masala Hot Roti Shoppe in Scarborough.'

'Yeah ah forgeh 'bout your Tobago connections. Why yuh doh cut down on de walking and spend more time at home with de wife or visit relatives? Maybe read a good book.'

Ravi shook his head. 'Nah. Ah doh want to learn anyting dat ah doh know. Doh dig no horrors with dis walking. Ah plan to cut it down soon. Meh lifestyle too strict. Since ah small ah grow up in a strict family. I is ah changed man. Now ah going to spend more time liming with meh friends.'

'So yuh still doing dat diploma programme at Cipriani College of Labour and Co-operative Studies?' Raphael checked his wristwatch. They had been walking for fifty minutes.

'Na ah drop out of it. It was taking too long and it would not have helped me with promotion at Dindial's Auto Supplies. Larse year ah play for de College in de Inter-Campus Tertiary Football competition. I was a defender on de team buh we place third. We play against ROYTEC, John Donaldson Techncal Institute, Hotel School, University of the West Indies, University of Southern Caribbean and University of Trinidad and Tobago.'

Raphael nodded. 'De odder day ah meet ah lecturer from de University of Southern Caribbean who was explaining how Communism is relevant for de world and dat de bourgeoisie will soon collapse. Yuh know it so easy to label someone or something as bourgeoisie. Some of dem university lecturers and odders like to talk and write 'bout how relevant Marxism and Socialism is to this world and all dem odder dead ideologies is to we society. Buh yet dey very materialistic and very much part of de capitalist system. Dey driving expensive cars, going for holidays abroad and living comfortable lifestyles and have not done anything for their beloved working class.'

'Dem is ah bunch of hypocrites,' replied Ravi. He coughed and then cleared his throat. 'Yes, is true wot yuh say, de always talking 'bout de working class dis and de working class dat. Dey love to talk and write of revolution or violence as a solution buh dey 'fraid to pick up a gun or be part of any radical movement. All ah dem is ah set of armchair critics who only collecting a salary and not making a real contribution to society. All dey could do is talk, talk, talk. De university supposed to provide some direction to the society.'

'Even worse is de fact dat some of dem lecturers geh tenure, promotion and win awards but dey hardly publish scholarly works. In dis country is who yuh know and who know you.'

'Yeah, dat is true all over de world. Some university people does geh promoted because dey regularly attending department and faculty meetings.'

There was a slight drizzle. Raphael looked concerned and wondered if Ravi intended to walk in the rain. He saw a piece of bottle nearby. He stopped, picked it up and threw it into the bushes.

'Come on let's turn around and head for home. Ah feel a few drops and it doh look like a passing cloud,' said Ravi.

Neighbours were surprised to see Ravi walking during the day without a hat, umbrella or sunglasses. He stopped after five months of this rigourous walking routine. He began complaining of vision problems and made an appointment with an optometrist at Ferriera's Optical.

After the appointment he returned home. Ravi sat on the couch and stared at the aquarium which had goldfishes and guppies. He glanced at the clock. Three more hours for Margaret to return home. He fell asleep on the couch.

Margaret returned and shook him. 'Honey wake up. You not going for your walk?'

Ravi slowly opened his eyes. He sat up and yawned. 'Sit down ah have some news for you.' He opened a box of Chuckwagon and bit into the fried chicken.

Margaret obediently sat next to him. She removed her silver bracelets and wristwatch.

He pointed at the sunglasses on the table. 'Ah went to the optician today and after some tests, she tell me dat ah have cataracts and will need surgery or I will go blind.' He took a small plastic fork and moved the fries in the box. He poured ketchup and mustard and on the fries.

'What! Is de smoking cause it?'

'No of course not. I stop smoking awhile now. She feel is all de walking in de sun.' Ravi yawned. 'Today ah also went by de chiropractor and he say de back pains and knee problems is a result of all de walking.' He paused and placed fries in his mouth. 'Do you know what I have to do?' He closed his eyes and began chewing.

'Yes dear, yuh need a tested and polarized sunglasses and have to buy a good pair of walking shoes. And you could walk in de early morning before the sun is up or late evening.'

'No, puh ah big X on yuh face.' He sighed and continued chewing. 'De only choice is ah have to immediately stop walking.' He opened his eyes and pointed to his chest and leg. 'Look at dese ugly scars. Is ah result of de surgery by a butcher who feel he is ah doctor. If yuh know how ah does feel shame when ah walking or go to de beach and people see it. For de rest of my life it will stay with me.'

'Buh walking good for de body. All de articles and dieticians say walking is good and....' Margaret raised her eyebrows. She opened her mouth to continue.

Ravi raised his hand to interrupt her advice. 'Wait. If yuh want me to walk den ah will geh blind and need a wheelchair. And if ah geh blind, yuh will have to leave yuh big wuk as a secretary at the Parts Depot in Longdenville and stay home and take care of me. Yuh want meh geh blind and cripple?'

Margaret nonchalantly responded, 'No dear.'

Ravi was glad that there was no argument. 'Good. Well now dat walking is finished. I need to start watching more television because soon ah might go blind if de cataracts not removed properly. Remember dis is Trinidad. Tomorrow ah going to install cable and start enjoying meh life. Ah also want to organize ah alumni association for former graduates of the St. Ann's Psychiatric Hospital.'

'Former graduates? You mean former patients?'

'Yeah whatever. All dem schools and universities have alumni groups and reunions, so ah don't see why former mentally ill people should not have reunions and alumni groups. You have ah problem with dat?'

'No dear. It's a wonderful idea.' Margaret shook her head. She wondered if he had a nervous breakdown.

'We need medical insurance. Dem public hospitals nasty and stink. And is poor treatment. When ah was dere ah heard dere was a woman who was supposed to have twins. De doctors tell she de twins is a boy and a gyul. De woman buy two cribs and plenty baby clothes. After ten months she realize something wrong so she went to de hosptial to deliver de twins. Yuh know wot happen?'

'Both or one of de twins died?'

Ravi shook his head. 'No. She wasn't pregnant all dat time. It was a huge tumour growing.'

'Oh, dat reminds me. Mrs. Pang had a boy while yuh was in de hospital.'

'She had a baby? asked Ravi. He laughed. 'She was always fat and nobody would ah know dat. Anyway, tomorrow is the Panorama semis so we have to leave home early.'

'Yes.' said Margaret. 'Ah eager to hear Tornadoes, Nu Tones, Trinidad All Stars and Invaders. I'm tired and going to sleep now.'

'Yeah ah will go later.' He pressed the remote and began changing television channels.

The shacks in Enterprise had Christmas decorations. Raphael and Juanita had placed flashing icicle lights outside their porch and along the edges of the roof. There were blinking lights on the mango and orange trees. Raphael felt there were too many lights. Inside their home, near the television there was a small Christmas tree and four Christmas stockings. Above the stockings were framed black and white photos of Marcus Garvey and Haile Selassie. Both pictures belonged to Raphael's father who had met Selassie during his visit to Trinidad. Under a faded photo was the caption 'Africa for the Africans.' The photograph of Selassie had the handwritten caption- 'Visit to Trinidad.'

Raphael was in the backyard cutting the grass with a cutlass. Marcus and Malcolm were playing football under the mango tree.

There was a knock on the door.

'Anybody home?' asked Ricardo. He fixed his collar and passed his hand over his hair. 'Man ah looking smooth.'

Juanita was nervous. She wondered if the police had come to question Raphael on the stolen items from Port-of-Spain or to break down the shack. She peeped through the curtains. She felt relieved. 'Yeah jus' a minute.' She patted her hair and adjusted her blouse. She cautiously opened the door. It was not the police. She welcomed him into her home.

Ricardo was a member of a newly opened church in Enterprise. He was distributing food hampers to needy residents. The visitor wiped his shoes on the mat and entered. He sat on the couch and folded his arms. 'Good morning marm.'

'Yuh lucky to ketch meh home. Ah was going to de market to buy some vegetables and fruits.'

'I'm giving food hampers to needy families.'

A serious expression came across Juanita's face. 'Doh take it de wrong way, buh we doh accept handouts. Go and give some needy people.'

'Marm…is not handouts. Is…charity.'

Juanita raised her eyebrows. She rolled her eyes upon hearing his explanation. 'Look arrong dis house. Yuh see any signs dat we is a charity case?'

Ricardo nervously glanced around the room. He saw a cabinet with expensive crystal vases. In the living room there were two large ceramic ornaments and on the walls had paintings by Boscoe Holder and Michel Cazabon. The family owned a colour television and two Beta video recorders. In one corner of the room was a stereo with two large speakers.

She pointed to two rugs in the hallway and a brass ornament on a table. 'Dem from India. Ah buy it larse month in de trade fair at Divali Nagar. And dem three paintings on de wall is all original.' She raised her arm. 'Yuh see dese two bracelets on meh hand…is real gold.'

He shook his head. 'Ah apologize for making de offer. Could I leave two bags of food so yuh could give someone who need it?' He wondered if she was involved in drug trafficking.

'Sure not a problem. Leave dem in de porch. Yuh face look familiar. Did we meet before?' She scratched her head. 'A party or fete?'

'No, I don't think so.' He rose from the couch and headed for the door.

She raised her hand and became excited. 'Wait. Yes, it was at two curfew parties in Marabella and Diego Martin!'

Ricardo was solemn. He adjusted his tie. 'Yes, ah did attend some curfew fetes during the coup. Buh dat was before ah became a born-again Christian.' He felt guilty and abruptly ended the conversation. 'Okay, bye.' He exited the gate and hurriedly left the neighbourhood.

'Mister you take care.'

Juanita departed for the kitchen. She checked the pots on the stove. The mixture of ochro, saltfish and rice needed more water. She took an enamel cup and filled it with water. She slowly poured the water into the pot. Fifteen minutes later there was another knock on the door. Juanita became frightened. She peeped through the window. It was Ricardo. She slowly opened the door. She saw the two bags of foodstuff which he had placed near a chair. 'Yes, did you forget something in de house?'

'Marm ah want to invite yuh familee to meh church. Service is on Sunday from 9am to 11am. De pastor will soon receive a big inheritance from a wealthy man in England who recently died and with de money he plans to build a real big church.'

Juanita would have politely refused the invitation but she felt guilty that Raphael had illegally taken items from stores in Port-of-Spain during the coup. During the past week she had also thought about the issue of life after death. She would have gone to a temple or mosque if someone had made the offer.

'Where it located?'

'Down de second road turn in Bhadase Street and walk right dong till yuh meet de crossroad. There used to be a mango tree. Yuh know it?'

'Yes.'

'Well it not dere again, somebody cut it down. Keep on walking and de church on de left. There was a parlour on de spot but it not dere again.'

Juanita listened and nodded. 'Ah feel ah know where de church is. Wha' time yuh say is service?'

'Is at nine.'

After he departed Juanita felt better. That evening she told Raphael her plans. 'Lissen, Sunday ah going to church. Some man invite meh to the new church up de road.'

Raphael nodded. 'Yeah okay. Maybe we can have our wedding dere. Ah have to go to a memorial service for meh aunt who had de funeral a few months ago. Ah will carry de chirren with me.'

On Sunday, Juanita arrived early at the church and sat near to Ricardo. He was glad she accepted the invitation. She was dressed in a white dress and a shiny, blue shawl was draped across her shoulders.

'My brothers and sisters, de world coming to an end. Repent sinners the end is near. If there are no trees, no forest and no jungles, then this will only mean one thing, that all the creatures of the forest- lagahoo, la diablesse and douen will be homeless!' shouted Pastor.

There were thirteen persons in attendance. The church had sixty chairs, a stage with a podium and two microphones. There were four bouquets of fresh flowers at the base of the podium. A large silver crucifix hung behind the podium. The floor was tiled with porcelain tiles and expensive purple drapes hung over each of the church's six windows. The choir had four women and each member wore a white gown. Two of the women were former Anglicans. Each choir member had heavy make-up on her face and pink or red lipstick. They smiled throughout the service. The services were video-taped and aired on TV6 on Sunday mornings.

'Amen!' shouted a member of the congregation.

'Preach on brother!! Hallelujah!' shouted Juanita. She jumped and clapped her hands.

Ricardo looked stunned. He motioned her to stop. 'Behave yuhself. Pastor fourteen security people doh bite nice.'

'Why he have so much security? He have more security dan de president!' She noticed that two of the guards were staring at her.

Ricardo did not reply. He listened attentively and regularly nodded whilst Pastor talked.

Juanita could not understand why Ricardo felt offended. Maybe this was because she was not a member of the church. She began to flip through the Bible. She did not own a Bible but felt it was important for her home to have at least one religious book. 'Ah want to borrow dis book for awhile. It sounding real good.'

She looked at the floral decorations near the podium and wondered which shop provided the flowers. She planned to tell Ricardo that she would be willing to supply flowers on a weekly basis.

'Okay, buh remember dat is de church Bible so doh forget de address.'

Pastor began to walk along the aisle as he preached. 'Plastics is the product of the devil. Plastics is not biodegradable. Burning plastics cause toxic fumes. Tings made of plastic like disposable diapers, car bumpers and false teeth are wicked. We need to start dividing our trash into biodegradable, natural, reusable. Of course don't do this on the Sabbath. All of we need to use more solar energy.'

'Praise be to God,' murmured an elderly man sitting behind Juanita.

'Too often I see de bad disposal of used car oils, styrofoam containers and paints. Who could forget de dump at Forres Park and the children with high levels of lead in their bloodstream who live in Demerara Road Settlement in Wallerfield?' Pastor wiped sweat from his forehead. He began to wildly wave a newspaper clipping. 'And some of you all might remember the illegal quarrying in the mountains and de brave lady who tried to stop it and had to place a gun at she head. And many might remember de former politician…who protested de paving of de savannah in town and had sand dumped on he legs. Pesticides are evil. Dey enter our body through respiration and it is absorbed by our skin. Farmers spray pesticides on food, we eat it and get sick and die. The farmers who use it also evil.'

Someone from the back shouted, 'Preach on Pastor!'

Pastor paused and drank a glass of water. 'All drivers who drive trucks, maxi-taxis and cars with broken exhaust pipes are going straight to hell. De cars produce black poisonous smoke which lead to lung cancer and other serious health problems like brain damage.'

Bhagwoutie, an old lady, began to tremble. Juanita continued reading the Bible and glanced at the old lady. She wondered if it was a case of epilepsy.

Bhagwoutie raised both hands in the air. 'Preach man! Preach! Yes Lawd, ah could feel de spirit.'

'Things such as air conditioners, refrigerators and deodorant and aerosol sprays are creations of Satan. These products release chlorofluorocarbons which deplete the ozone layer and allow dangerous ultraviolet radiation. Is dat causing de flooding in San Juan and in de south.' Pastor paused to wipe the sweat that was trickling down his face.

Juanita thought about the leaking refrigerator in her home. She wondered if it had an evil spirit.

Bhagwoutie fell to the ground and started rolling and shaking. She shouted, 'Amen! Praise be to God!' Juanita looked alarmed. Ricardo did not seem concerned with the reaction of the old lady. A middle aged man at the back of the church fell to the ground and began speaking gibberish.

Ricardo turned to Juanita and smiled. 'He speaking in tongues and she have de Holy Spirit. When it came to the Bible and the environment, I was confoffled and used to buss meh brains trying to find the link. Buh ah can see clearly now thanks to Pastor. The Pastor is not an ordinary person. He is now an apostle.'

Juanita nodded. She was not planning to roll on the ground because it would dirty her white clothes.

'The world is coming to an end!' Pastor began to wave the Bible in the air. He had the biggest Bible in the church. 'Yes is ALL true! Dat is de Lawd's word to man. We have to recycle, reuse, reduce and recover. The signs mentioned in the Book of Revelation in the New Testament are obvious.' He slammed the Bible on the podium. 'Dat is all folks. Next week we celebrate de Passover. Remember de Bible is de truth. Yes is ALL true! Dat is de Lawd's word to man...and woman. Crick, crack monkey break he back on a piece of pommerac.'

Juanita thought it was an odd way to end his sermon. She remembered those words being used by a local storyteller- Paul Keens Douglas.

The choir sang two songs. The singers were offkey and the organist did not seem to know the music. Nobody in the choir seemed to care about singing in harmony. There were seven microphones among the choir members. The choir seemed like a showpiece of the church. The choir conductor was a middle aged man who furiously waved his hand at the four women. He acted as if he was directing a large symphony orchestra at the Royal Albert Hall in England.

Pastor looked at the congregation. 'Anyone would like to be saved or healed?' He looked at Juanita.

She murmured, 'Ah good today.'

Pastor looked at her and motioned for her to come forward.

'Ah good today. Buh yuh think yuh would be able to heal meh fadda broko foot, meh cousin cleft lip and meh mudder fini hand? People does only call meh fadda – hop and drop.'

Pastor shook his head. 'NO. Ah not healing dem kinda people today.'

Juanita raised her voice, 'What? How yuh mean not today? Doh leh meh flare up like a kerosene stove.'

'Ah doh deal with dem kinda ting. De pastor four houses down de road does specialize in healing cripple people, retarded people, fini hand and broko foot.' He turned and headed for the podium.

Juanita was stunned. She whispered to Ricardo, 'Dat is high class nonsense! Like he is a specialist healer?'

Ricardo placed his finger on his lips. 'SHHH. Learn to handle yuh stories. Yes he is a specialist, ent it have doctors and lawyers who is specialist?'

'Yeah.'

'So behave yuhself.' He pointed at the security personnel.

An old man kneeled in front of Pastor and began to beat his chest. 'I am a

sinner, I squat on the hillside in Maracas. I responsible for some bush fires, slash

and burn agriculture and deforestation. Now I want to protect de environment!'

Pastor placed his hand on the man's head. 'Go in peace, your sins are forgiven. Squat no more. Everyone let us pray.'

Everyone closed their eyes and bent their heads.

'Dear God, we all know dat jump high or jump low…we have a duty to preserve and protect de environment. Help we to make this country a beautiful Garden of Eden. Dear God, help those persons who are affected by pollution such as those in China and India. We ask you to stop acid rain, help reduce greenhouse gases. Help we to save de Amazon. Look at de flooding in our country! Is pollution cause dat. As we see in Genesis, you took only six entire days to create the world.' Pastor pounded the podium with his right hand. 'Lawd help we to stop polluting dis blessed paradise and de Amazon. Amen. Amen.'

Juanita murmured to herself, 'Dis is blasphemy. Buh wha' de jail is dis? Jus' so pollution go stop? Dis pastor only bumping he gums.'

Ricardo said, 'Dat man could pray yes. Ah doing my part. Look jus' larse week ah buy two books from Amazon and it ship to my house in four days. We certainly need to protect Amazon.'

Juanita touched his shoulder. 'Pastor talking 'bout de Amazon forest not de internet.'

Ricardo recoiled and looked at her. He dusted his shoulder. 'Doh ever touch me. Yuh will give meh bad energy and ah will geh sick. You could pass on bad spirits to me.'

She profusely apologized.

'If allyuh know wha' good for allyuh, den allyuh would not pollute. WE will now give thanks to God for all that he has provided for us.' Pastor then passed two offering plates to the elders. They began passing the plate to each person. The plates were returned to the elders. Each plate was overflowing with twenty and one hundred dollar bills.

Bhagwoutie returned to her seat near to Juanita. Her white dress was dirty and she was sweaty. 'I is ah reformed woman, ah born again, conservationist. I going to stop squatting and be against quarrying that taking place on the hills and mountains. Ah will also stop passing meh mouth on people.'

'Good for you,' said Juanita. She felt this church was a cult and was glad the service would soon end.

Bhagwoutie wiped the sweat that was trickling down her forehead. 'Yes. After that workout ah doh need to go in ah gym. Dis is meh weekly exercise.'

An untidy and foul-smelling vagrant entered the church. His name was Thomas and he tightly clutched an old plastic bag. He stood at the back of the church and watched the proceedings. Juanita noticed the intruder in the building. After five minutes two security guards escorted the vagrant outside the church.

Ricardo explained the situation to Juanita. 'He does always come in dis church on Sunday morning to beg for money and food. Leh him go by de Anglican or Methodist Church up de road. Our church is for de wealthy and clean people. Jesus was no beggar and always talked about de need to be clean.'

Juanita was ignorant of the teachings of Jesus and decided not to respond to his comment. She wanted to demonstrate her concern for the environment. 'You should be like me. I am a member of the National Reforestation and Watershed Rehabilitation Programme.'

Ricardo looked surprised. 'Who? You? Nah nah. Dat cyah be true. Yuh always pelting empty food boxes and beer bottles out of your car window.'

Juanita shook her head. 'Not me. Must be many years ago…before I was saved. I is also a member of the group Citizens in Action to Restore the Environment which is organized by the First Citizens Bank. For World

Environment Day I does help de Colonial Life Insurance Company to distribute thousands of free trees which they give away each year. And ah going to join that radical group Fishermen and Friends of the Sea.'

Ricardo smiled. 'Ah yes…dat insurance company…one of dese days dat company go buss. Dey expanding too farse and investing people money in ah setta tings dat not profitable. De safest place for my money is Republic, Scotia or Royal banks or under meh bed mattress.'

Pastor stood at the back of the church. He shook everyone's hands.

The choir director asked Pastor, 'Are we singing for the church's anniversary service next month?'

'No.'

'Why? We have been rehearsing two songs.'

Pastor was serious. 'You all know you cannot sing properly.'

The dejected choir leader meekly exited the church. She was deeply offended and relayed the message to choir members at the weekly practice session.

Ricardo pointed to Pastor and said to Juanita, 'Is true wot people say. If God cyah come he does send a man.'

'Oh my God look how late it is. Ah have to go home and fix up de house. Raphael cousin from Sweden coming to visit.'

'It have Rasta in Sweden?' asked Ricardo.

'He is not a Rasta. Buh it have Rastas all over de world.'

Juanita walked to the main road and waited for a maxi taxi. She met the old lady who sat nearby to her during the service. The old lady was waiting for a taxi. She stared at the gold chain around the old lady's neck and the two gold bracelets on her arm. She was convinced the old lady was wealthy. 'Hello marm. My name is Juanita. Do you need anyone to help clean yuh house?'

Bhagwoutie was apprehensive. 'Well…ah…ah need somebody to do a lil mopping and cobwebing every month.' She nervously touched the two gold bracelets.

Juanita noticed the old lady's gold earrings and became excited. 'Great. Do you live far from here? Do you have children? Is your husband still alive?' She took a piece of paper from her purse and began writing.

'Meh husband dead larse year. Meh two chirren big and married. Dey living abroad. Ah living in de white and green house at de corner of Jerry Junction in Waterloo.'

'Yeah ah know dat house. Here dis is my telephone number.' Juanita took another piece of paper from her purse. 'Here put yuh number here.' She thought about the sermon and the choir members. She would not return to the church. She saw a maxi taxi approaching. She waved and it stopped.

The driver's assistant shouted, 'Two for Curepe! Two for Curepe! Darling yuh goin' Curepe?'

Juanita replied. 'Ah going Warrenville.'

The assistant nodded. He opened the door and offered her a seat. Here come and sit here. He watched her and shook his head with approval, then smiled and rubbed his hands.

'Thanks.'

'So sweetness yuh have a name? Or do people jus' call yuh beautiful.'

Juanita stared outside the window. 'Ah is ah married woman.'

'Reds, me eh arsk yuh all dat. Not often ah get to meet nice people who intelligent, does dress good and pretty. So where yuh from?'

The other passengers smiled. They were accustomed to these remarks and knew the outcome. It was a hot day and the driver adjusted the controls of the air conditioner. He squinted as glare was intolerable.

She looked upwards and rolled her eyes. 'Japan.' A passenger laughed.

He smiled and detected the sarcasm. 'Really? What a coincidence! Ah from nearby China. So you travel dis route often?'

'Yes, only when I want to be harassed by dotish lil boys. Lissen tell de driver dat ah taking it by the traffic lights.'

'Drive stop by de lights.' The assistant looked at Juanita. 'Yuh leavin' already? Ah could geh yuh cell number to keep in touch?' He had a broad grin and did not expect an answer.

'Sure is 665-NO. Give me a call this evening and we could arrange to get to know each other a little better.'

He was taken aback and immediately typed her number on his cellphone. 'Sure ah would like dat.' He winked at her.

She was surprised that the assistant believed that was her real number. At the traffic lights, the maxi taxi stopped. Juanita paid the assistant and departed the vehicle.

'Buh ah find dis maxi tout well fresh with heself,' said Margaret. She wore a business suit and had a briefcase on her lap.

Patty replied, 'Gyul dat is ah normal ting in dese maxis and taxis. Some of de young boys like de job because de have ah set of opportunities to pick up young tings.'

The assistant asked Patty, 'So darling, where yuh from? Ah know somebody looking like you could never be single. Gimme yuh name and number na.'

Patty looked surprised. 'My name and number? When cock get teeth! Yuh too fresh with yuhself. Ah married and have four chirren.' She turned her face and stared outside.

'Ah didn't ask yuh all dat. Cool yuhself. Ah jus' trying to make a lil conversation.

Wha' 'bout yuh friend. She single?'

'Leave us alone,' said Margaret, 'We have HIV and AIDS.'

'So yuh cutting style on me!! Jus' because yuh have 2 degrees- HIV and AIDS, doesn't mean yuh have to treat me so. Doh feel big because yuh have Bachelor and Masters degrees. One of dese days ah go get a degree. Everybody getting degrees like MBA dese days. And because dey have degree dey does treat me de same way- as if I stupid.'

Patty raised her voice. 'Yuh have to be joking! Yuh mean in dis day and age yuh doh know what AIDS and HIV mean? All yuh man doh care. Yuh only care about sex and not care about the disease.'

'You feel as if yuh know everything. Two of you eh sneezing and coughing so ah know yuh healthy. Ah know how to deal with disease. Look larse week ah had de flu buh after taking two multisymptom Panadol tablets and using Vicks ah was good de next day.'

The maxi taxi stopped at the Curepe junction. Both Patty and Margaret paid the fares and quickly disembarked from the vehicle.

Chapter 4
Surviving

Juanita reached her home. She saw Marcus and his friends playing football. She was disappointed and felt her son possessed talent and potential and should be playing on the national team instead of wasting time playing football on the road. She decided that next week she would have a serious chat with him regarding his future.

Juanita called her two daughters. Cleopatra and Harriet were fifteen and eighteen years old. They were brown skinned and disliked their broad noses, curly hair and thick lips. Both hated their lowly status in society and regularly discussed their dreams of becoming wealthy and going abroad for annual holidays, possessing a bigger home and expensive cars.

Their mother had a serious expression. 'Ah have some good news. Today at church ah met ah old woman name Miss Bhagwoutie. She living alone in a big house! She chirren big and living abroad.' She stared at her daughters and patiently waited for a response. They listened attentively and knew this was their next victim.

'Harriet since yuh older ah want yuh to go every Sunday and meet dis woman by she house and get a taxi or maxi to carry she to church. See if she want someone to cook for her. If she say yes den we go cook a few meals for her.'

 Harriet nodded.

Juanita pointed to Cleopatra. 'Ah want you to go and clean she house every month. Tell she dat yuh could clean it every week. See if she want anybody to comb her hair or cut her nails.'

'Buh mommy why we taking care of Miss Bhagwoutie and we ignoring we own sick grandparents who need help?'

Juanita remained quiet and stared at one of the ornaments on the table. 'In church ah made a promise to Miss Bhagwoutie to take care of she for de rest of she life. Ah intend to keep dat promise. Ah want all yuh to be caring. Is ah Christian thing to take care of ole people. Is in de Bible.'

Both Harriet and Cleopatra nodded.

Marcus would usually spend his afternoons playing football on the road with Quasee, Anil, Ming, Nigel, Gabriel, Kervin and Aboud. Quasee was stocky and of Chinese descent and Anil was slim and of Indian descent. Nigel and Ming were slim. Nigel was of Indigenous origin whilst Ming was of Portuguese and Spanish heritage. Gabriel, short-tempered and fat, was of French and Jewish ancestry. Kervin was tall, amiable and known as Dougla because he was mixed with African and Indian heritage. Aboud's parents were from Lebanon and Syria. These teenagers were neighbours and childhood friends of Marcus.

Six years ago, Anil and Aboud played in the Atlantic LNG National Primary Schools Football League Competition. Anil also represented Trinidad and Tobago in the Under-13 and Under-15 football teams. Such involvement earned him respect from the other boys.

Some of the players had their strengths and weaknesses. Nigel was good at penalty kicks and Ming was a good defender. Kervin was a playmaker and not skilled in shooting at the goal. Marcus was a forward who had a powerful left foot. Gabriel was always boasting of how many goals he scored at his secondary school. However, he never scored during small goal football on the road. He always wanted to make a solo run and receive praise for scoring goals.

Quasee's nickname was 'Dribbler' because he would easily avoid opposing players who sought to tackle him and get possession of the ball. He did not like school. Whenever he was at home, he would spend many hours in his backyard learning to control the football. He would be raising the ball, balancing it on his forehead and behind his neck. These tricks would be used to impress the other players during the evening games.

One Sunday night Marcus entered his home and was confronted by an angry Juanita.

'Yuh wasting yuh time and yuh life. Whole day yuh kicking ball or bouncing ball. Why yuh doh apply to de bank or go university and do a degree?'

'Mommy de small goal on de road and de basketball is for fun.' He wiped his sweaty face.

'You have two A Level subjects and yuh selling by de roadside.' Juanita placed her hand on her waist. 'Which woman go ever want to settle down with yuh?' She felt Marcus had wasted a promising football career. He was a former captain of the Naparima College team which defeated Queen's Royal College in the finals of the Coca Cola National Intercol competition.

Marcus opened the refrigerator and poured cold water into a glass. 'Doh worry is jus' a lil sweat on evenings.' He did not want to argue with his mother. He stared at the ceiling and saw cobwebs with three small spiders.

'Yuh need to be financially independent.'

He refilled the glass. 'Buh I am making some money. And ah doing something ah like.' It was nine o'clock. He wanted to end the discussion and get in the shower. A mosquito landed on his arm an dhe quickly killed it.

'Marcus ah want some help with de flower shop.' Juanita had a flustered look.

He was hesitant to agree. 'Wha' exactly you want meh do?'

'Ah need yuh to check the death announcements in de papers every day and call de people to see if dey need flowers. If dey say no try to convince dem of de need for flowers.'

Marcus nodded. It seemed simple. He looked at the spiders searching the cobwebs for a victim. The spiders reminded him of his mother who was searching for customers.

'Ah also need yuh to regularly visit the cemetery and bring de fresh flowers from de graves for me to use for arrangements.'

'WHAT?' shouted Marcus. 'Yuh want meh take flowers from people graves? Mammy yuh eh feel dem evil spirit go hold meh.'

Juanita smiled. 'Son, things hard. We need de money. Ah not asking yuh to murder nobody. And dem dead people will be glad to know de flowers helping out poor people like us. Yuh sisters already helping as much as dey could.'

Marcus tried to find logic in his mother's explanation. Eventually he grudgingly agreed.

The money earned from the flower shop was sufficient to pay for the rent, groceries and transport. Customers purchased flowers for anniversaries, birthdays, Valentine's Day, weddings and funerals.

Raphael earned a meagre salary. During the evenings, he operated a fruit stall near the Kentucky Fried Chicken outlet in Chaguanas. He would jokingly tell his customers that Colonel Sanders, the founder of Kentucky Fried Chicken, is the Black man's hero for discovering fried chicken. He sold bananas and mangoes but during December there were crates of apples and grapes for customers. He also had two small bottles with dinner mints and cigarettes.

Ten year after the coup of 1990, the life of Marcus had changed. Raphael was sentenced to six years in jail for possession of marijuana. Marcus still assisted in the family's flower shop. He had purchased an Almera car and used it to deliver flowers to customers and also to transport stolen flowers from the cemeteries to the flower shop. After six months, thieves stole his car. Bandits and kidnappers in Trinidad were stealing Almera and Sunny cars.

He often saw death notices which stated- 'No flowers by request. A collection will be picked up.' Some notices provided the name of a charity where money could be sent. This did not deter Marcus from calling the bereaved family. He would check in the telephone directory and dialled telephone numbers until he found the grieving relatives. He would ask if they needed a wreath on the coffin or flowers in the church.

Anil visited the shop to buy flowers for a birthday party. He wore a track suit, gold chain and on his left hand were gold rings on his fingers. Around his neck were headphones.

He asked, 'So how de flower shop?'

Marcus smiled. 'Boy, money flowing like water. Only a rumshop could make more money dan dis flower shop.'

Marcus was a member of the Tropical Angel Harps Steelband located on Bhadase Street in Enterprise. One evening he was with other players rehearsing for a performance when he saw Matilda walking along the sidewalk. She was slim, tall and had shoulder-length hair. She wore a tight red blouse and a short cream skirt.

Marcus and his friends saw her. They stopped rehearsing on the steelpans.

Marcus was the first to react. 'Pssss…familee, psssss…famileee. Yuh sweet like a jub-jub, psssss…familee, psssss…familee.' He left his friends and began walking towards her. 'Babe, how yuh treating a brudder so?'

Matilda stopped and looked at Marcus and the other steelpan players. 'Not interested. I'm looking for a decent person.'

Marcus was surprised that she had stopped to respond. Usually girls and women ignored the comments of the steelband players. 'I man decent.'

'Yeah right. Yuh too fresh. Ah looking for a boyfren' who willing to respect me and not only want to sleep with me and use me.'

'Sorry babes buh nobody like dat exists.' He walked closer to her. 'Dem extinct like de dinosaurs. So babes, wha' is yuh name?'

'Matilda.'

'Now dat is ah pretty name. Reminds me of fresh hops from ah bakery. Just hearing it make me feel hungry.'

She began to walk away.

Marcus followed her. He wondered if he was too intimidating. 'Wait na. So yuh have any future plans for dis imaginary so-called boyfren?' He slowed his pace.

Matilda continued walking and responded, 'Ah want a future husband who has values, morals and a sound education.'

'Gyul dem kind people doh exist, dey imaginary like Santa Claus and de Tooth Fairy.'

'Doh talk about teeth because yuh look like yuh need to see two good dentists.

How come some of yuh teeth crooked and de teeth in front big like lumber from ah hardware.' It was windy and she adjusted her hat.

He waved his hands and pointed to his mouth. 'Meh teeth is minor. De rest of my body good.' He wanted to end the conversation and return home but there was something intriguing about this girl.

'Good for what? De cemetery? Yuh teeth yellow like a ripe banana. Like you and toothbrush is no friend.' She quickened her pace.

Marcus was persistent. 'Darling, leave meh teeth alone. Dis is about mankind and you.' He felt awkward defending his physical looks. The sky was blue with few clouds. It was a good day to visit the beach or play cricket.

She glanced at him. 'I cyah handle some marasme boy who feel dat having his pants low down, a fat chain round he neck, some cheap lyrics and earrings in he ears, is being a man.' This was not true because she loved men who wore chain and earrings.

'Yuh tink I is ah big duncey head. Ah sure yuh boyfriends belong in a horror house and have bad hairstyles.' He quickened his pace and sensed that she was attracted to him. 'Yuh jealous.' He wiped the beads of sweat that were trickling down his temples.

'Me jealous? Why ah go be jealous of somebody who does look like a lollipop with a big head and a small body?' She quickened her pace and hoped the harassment would soon end. Her body language indicated she did not want to continue the conversation. 'Lissen, mango head boy…yuh breath smelling really bad…like a cesspit.'

'Doh insult me…ah does use Colgate and Pepsodent.'

She smiled. 'Really? How often once a month?' His witty responses to her insults had an effect. She was beginning to believe that it might be a harmless attraction or infatuation.

He grinned. 'Three times a day.'

'Three times? For brushing? Like yuh does use it to polish yuh shoes three times a day.' She felt a shortness of breath. This was partly due to her being overweight.

He placed his hand by his mouth and exhaled. 'It not that bad.'

'Yeah right!' She giggled. 'Yuh have to use something a little stronger.'

'Like what?' asked Marcus. 'Bleach or Kerosene?'

Matilda smiled. 'Yuh know, yuh have some real lyrics. Well, don't you have something to do…like playing hopscotch with yuh friends from kindergarten.' She passed Vierra Street and thought about the money she owed her cousin. It was one thousand dollars and she had borrowed it two years ago.

She stopped and checked the road for incoming vehicles. The light turned red and the vehicles stopped. She crossed the road. Marcus wondered if he should also cross the road. She glanced at him and he decided to continue following her.

'Yuh too sour.' Marcus stuck his hand in his pocket. 'Yuh want a Diana Powermint? Anytime ah have problems ah does suck one.' It was an awkward conversation that both seemed to enjoy.

She shook her head. 'I see that yuh too poor to afford a psychiatrist or psychologist so yuh have to be happy with a mint.'

'Gyul, come na, everybody doing it. We have de space, place and de right mood. Ah could give yuh all dat yuh odder boyfriends promise. How yuh so grumpy like yuh want some Zentel tablets to clear up de worms in yuh belly?'

Matilda's tone was defensive. 'What you know about parasites? Dem worms must have more morals and a higher IQ than you. Like you does get yuh education from de television or de writing from behind a cornflakes box?' She opened her purse and checked for her sunglasses. It was not there and she was disappointed.

'No, actually from BET, MTV and VH1 on cable tv.'

'That explains a lot.' She had a look of contempt. 'Ah really hope yuh change yuh ways and stop feeling as if you is God's gift to women.'

'Ah will think 'bout it. Wha' yuh saying making some sense.'

She pointed to a flat house with palm trees and lilies at the entrance. 'I live here.

Yuh playing de fool, ah wasting my breath on you. And, remember find ah good woman and settle down. But I determined ah go live differently and make a difference. Ah go save myself till ah get married and remain faithful.'

She opened the gate of her home and began searching for the house keys. Marcus left and headed for his home. There was a feeling of disappointment but he was glad that she had the conversation with him. He entered his yard and looked at the damaged wheel on the box cart.

Two years had passed and Marcus still thought about Matilda. His community had changed. Squatters now had cable television, cellphones, swimming pools and luxury cars such as Audi, BMW, Mercedes Benz and Rolls Royce. There were fewer murders, robberies and kidnappings in Enterprise. This was partly due to the Neighbourhood Watch formed by concerned residents. Some of the residents suggested that Enterprise become a gated community. Others objected saying that this would create an elitist community. The residents decided to protest against the failure of councillors and politicians to repair the deplorable roads in their neighbourhood. Sometimes an old tyre or piece of wood was placed in a large pothole as a warning for drivers and pedestrians during the night. Residents used broken bricks to fill the potholes but this damaged tyres. It was an unwritten rule that roads would be paved only before the national and local elections.

One morning, the residents decided to block the roads with old stoves, mattresses, pieces of wood and iron, and old couches. They lit the debris and called the television and newspapers to report on their protest. Three residents held placards which stated- 'Worst roads in T+T', 'We MP forget we' and 'No road no vote.' When the police and firemen arrived at the scene the residents shouted expletives and refused to remove the smouldering items from the road. This method of protest was used by rural residents and poorer sections of the working class to get improved social services and also demand justice. It was street justice and invoked whenever the government ignored their problems.

Anil, Marcus, Ming and Aboud decided to shift their game to another street. This new location was away from the protest and it also offered them more space as the street was wider and recently paved.

The football eventually began to shred. The game would sometimes be temporarily halted as players removed parts of the shredding ball.

Anil clutched it and began to rip a piece of the outer covering of the ball. 'Dis ball well serve us long.'

Ming laughed. 'Yes, Spalding is a good brand. Dat ball take on gravel, pitch, mud and sand. All yuh see daddy on television larse night in de protest?'

'Yes, and meh sister was in de *Newsday* today,' said Anil.

Whenever the ball became soft, Marcus would carry it to the nearby bicycle repair shop to be inflated. The owner would not charge him for this service. One day the football became defalted but the repairman felt it was too old and could not be repaired. The boys decided to use a small, soft basketball. This was their football for the next six months.

Broken stones, pieces of bricks or shoes of players would be placed on both ends of the short street to represent goalposts. There would be cursing and shouting as the boys argued over claims that a goal was scored. It was sometimes difficult to decide if the ball had passed between the stones or over one of the shoes. On one occasion empty glass bottles were used as goalposts but this soon stopped after a car ran over one of the bottles and the irate driver threatened to report the boys to the police. On the field, during the rainy season, the goalposts comprised short sticks or broken tree branches which were firmly stuck into the soft mud. The boys enjoyed playing barefooted on a muddy field as it meant an opportunity for a sliding tackle.

Neighbours were not pleased with their noisy presence and often called the police. On Easter Sunday, two police cars arrived with flashing lights and sirens. The occupants of the car were heavily armed. The boys stopped playing.

A burly police officer with riot gear emerged from the car. He had a sophisticated semi-automatic gun. He looked at the boys and grimaced. 'Boys I've been receiving complaints from neighbours that you all are cursing and shouting late at night and disturbing the peace.' Other officers emerged from the cars. They wore bullet-proof vests and had guns.

Marcus and the other boys sought to explain the situation. Marcus casually dismissed the complaints, 'Doh worry wit dem one or two neighbours who complaining. Dem feel dey own de whole of de community. Who own de streets? Who own de main roads?'

'The government,' replied a policeman.

Aboud nodded. 'Exactly. And we too young to vote buh is we parents who vote for de government. We is not causing any traffic or tiefing, or murdering anybody. Officer, we juss getting a lil sweat on evenings. Doh dig no horrors.'

'We doh have a savannah to play in and thus we are forced to be here,' added Ming. The policemen were serious. He was trembling and stood near a hibiscus hedge. 'Some of we unemployed.'

The police officers listened intently. One advised, 'Okay consider this a warning. Ah don't want any incidents to erupt.' He returned to his vehicle and both cars sped away. For the next three years, the boys would play football on the road and the neighbours did not complain.

During the nights Marcus would sell oysters under a makeshift shed by the roadside near his home. It was peaceful and relaxing. He needed to unwind from the stress of stealing flowers from the cemeteries. There was a small table and wooden chair in the shed. On the table were bottles of ketchup, water, salt, lime juice and pepper sauce. At the corner of the table was a flambeau which comprised a small bottle with kerosene and a lighted wick. At the side of the shed was a heap of oyster shells.

Jessika lived three houses from Marcus. Ten years ago she was a classmate of Marcus at Longdenville Presbyterian Primary School.

'Marcus hear na, it have a nice preacher man in we church, He could sing some nice carols—like Away in Manger and Joy to the World. He talking and preaching and ting, about the world ending, yes it going to end. Buh doh worry, he say dat de good people living long and forever in heaven.'

'What is de name of de church?'

Jessika coughed and then spat in the nearby drain.

'Jess doh spit arong here. It not hygienic.' Marcus had a serious look. He looked around to see if anyone noticed. He did not want his sales to be affected by her action.

'Sorry 'bout dat. People does call de church Tabernacle of Believers, the Saved and Born Again Brethren. But it have ah odder name ah cyah remember. Anyways, dem is a breakaway faction from Jesus is de Light Ministries. It somewhere in Guanapo and dat was de church some dotish politician had given ah setta taxpayers money to build. They planning to open a branch right here in Enterprise. Anyway, hear ting about Janet. She is a member of de church and became a prophetess and now she advising politicians and businessmen. Whole day she prophesying buh she cyah tell meh de winning lottery numbers. She didn't even know she own son was on drugs. She so damn farse always minding meh business…and…wait….What I was telling you larse week about meh big problem?'

Marcus's disposition changed when he heard the name of the church. 'Ah know dat church. Ent yesterday yuh was telling meh something 'bout yuh sista horning she husband with somebody from dat church?'

Jessika looked embarrassed. 'Dat is true talk but no, is not dat church. Is annuder church.' She stared at the ground for a few seconds. 'Yes, it now come back to me, is about de little ole dawg that telling everybody she is meh daughter! Lissen give meh some oysters and puh plenty pepper.'

Jessika took a small, black purse from her bosom and gave five dollars to Marcus. 'Tell yuh fadda dat Barbara in hospital again. Ah hear she travelling.'

Persons would use the term 'travelling' to describe persons who were dying. Marcus smiled. 'Ah didn't know she gone dere again. She always in and out de hospital and forever on deathbed.'

'Yes. All de chirren from England and Canada come down to visit them. Margaret have she funeral clothes sunning out. She tell de chirren to put up a tent tonight for de wake.'

He accepted the money and gave a glass of seasoned oysters to Jessika.

Jessika stirred the drink with her index finger. 'Aye, meh neighbour brave yes and eh have no shame. She did tief two turkey from Hi Lo grocery yesterday…and give meh one so ah forgive she. How a young gyul only fifteen years stealing already?' She took a mouthful and began chewing.

Dorothy, Jessika's neighbour, slowly approached them. She was returning from the grocery with six plastic bags filled with food.

Jessika waved her hand. 'Good night. Hear na Miss Dorothy you hear anything about de lazy husband of yours?' She emptied the contents of the glass into her mouth.

Dorothy stopped near the stall. She placed the bags on the ground. 'Good night Marcus and Jessika.'

'Good night Miss Dorothy,' responded Marcus. 'How de husband?'

'He is ah waste of time, sleeping whole day and since July playing parang CDs and waking up late at night watching cable tv.'

Jessika placed both hands on her waist. 'Is true- a leopard never change he spots. If mine leave ah real 'appy. Why all dem man in Enterprise and Endeavour so damn lazy?'

'Young man how de flower shop?' asked Dorothy. She squinted.

'A bit slow dese days. Ah not sure why.'

Dorothy grabbed three of her bags. 'Yuh have to start selling geera pork or gyros and sales go be high like in de flower shop.' She felt Marcus was gay because he sold flowers. She believed that men should not be florists, secretaries or hairdressers.

Marcus laughed. 'I go do it next year if Jah spare life.' He had a cane row hairstyle. He scratched under his left arm. 'I want to start selling oil down and bake and shark.'

'Yuh hear dat the church up de road close down?' Dorothy gently massaged her left knee. She believed the rainy weather affected her arthritis.

'No. Buh meh mudder used to go to it. Wha' happen?'

Dorothy cleared her throat. 'Well de pastor suddenly believe in evolution and he say de story about Adam and Eve in de Old Testament is not the beginning of de world. He writing a prequel to Genesis. He say it had life on de earth before Adam.'

'A prequel?' asked Marcus.

'Yes, like how some of dem Hollywood movies have prequels...about how tings originate. Pastor say God speak to him and tell him to do de prequel and he mentioning dinosaurs and cavemen. Yes, he going to mention dat God did create evolution and dinosaurs and also aliens and UFOs. De church did have odder minor problems such as de treasurer who did steal ah set of money.'

Marcus shook his head and continued to clean an oyster. 'Black people cyah run ah business. Not even a small parlour. He writing prequel to de Bible?'

'Not jus dat. He say dat God tell him to add five more commandments. Dis is because it have more sins and new sins in de world so he adding them to a new version of de Bible.'

Marcus had a serious expression. 'More commandments? Since when God does speak to Black people? By de way ah hear Ravi drop a big lawsuit on Pastor.'

'Yes boy!' Dorothy laughed. 'Pastor could not heal his cataracts and he carry Pastor to court. De case might land up in de Privy Council in London.' She grabbed the rest of her bags and began to depart. 'Wait! Before I go, leh me tell you 'bout dat hard-head girl who is a cashier in Ramphalie Grocery in Endeavour. Yuh know she is meh outside chile?'

Jessika looked shocked. 'Doh make joke. Anyway dat is small ting. Lissen, ah saving money to buy some furniture from American Stores dis Chrissmass.'

Marcus was surprised. 'Buh ent larse month yuh buy a five piece living room furniture, a computer, three stoves, fridge and two television sets on hire purchase from Standards and Courts?'

'Yes, buh de fat lady…Mahendee Spice in de television advertisement

who does shake she tail and say 'nothing, nothing down' also say that customers getting ah free ham, den ah realize dis was ah real deal.'

Dorothy laughed. It was a loud laugh and her entire body shook. 'Lordy Miss Claudy and she million chirren! Why yuh don't jus' buy hams instead of all dat wasteful spending on appliances?'

Jessika glanced at Dorothy who had twisted her mouth sideways. She felt Dorothy was trying to humiliate her in the presence of Marcus. Dorothy did not like the friendship between Jessika and Marcus. Dorothy always felt Marcus was an ideal husband for her daughter, Ierene.

'Miss Dorothy, yuh feel yuh too bright. Chrissmass means to have a good time and spend money left, right and centre.'

Dorothy looked upwards. 'Well Lord look at meh troubles! Doh tell meh dat is a drizzle ah feel. Yuh mean it going and start to rain now.' She returned under the safety of the stall. 'Anyway Britney studying boyfren whole day in school. I fed up and damn vex. For Chrissmass de boyfren planning to give her an engagement ring. Buh if he do dat ah go beat him till he bawl like ah goat getting killed. Anyway, I tole you 'bout de latest news?'

'No. Tell meh later, ah have to go and put ah weave in meh hair. We go continue the talk tomorrow Marcus.' Jessika hurried along the street.

Marcus replied. 'Right Jess.'

Dorothy was agitated. 'Marcus is okay if ah stay here till de drizzle stop?'

'Yeah.'

Dorothy adjusted her green blouse and long skirt. 'Dis is de worse Easter ah ever spend. Me mudder, sick with kidneys and larse week she geh a stroke and heart attack. She deading slowly.'

Marcus threw a shell in the pile on the ground. 'Ent she was sick for de larse ten years? She leaving anything in she will for yuh?'

'Nine years to be exact. She deadin. Of course she giving me plenty things in she will, cause is only I did care.' She thumped her chest. 'She leaving meh de ole Chrissmass tree and de family Bible. Not dat me doh care or me cold-hearted but all de suffring and ting not de best ting in this place. Is better to die now dan take on all de kidnappings and murders in dis country.'

'Dis place is Babylon.' Marcus had always heard his father use the phrase and he began using it anytime someone mentioned the country's social problems.

'Amen, brudder that's right. Eh eh look big head Bally coming dong de road!' said Dorothy.

Bally wore a faded jeans and black high-top Puma sneakers. His shirt was unbuttoned at the top to expose a gold chain and hairy chest. He was the typical Caribbean macho male. He wore a shrivelled Santa Claus hat. There were blinking lights permanently affixed to his home's doorway and icicle lights were on the roof's edge. From September to December, he would adorn his five fruit trees with coloured, blinking lights. Throughout the year parang, Christmas hymns and carols would be blasting from five loudspeakers. Everyone in Enterprise regarded him as eccentric. 'Easter Greetings everybody. Marcus give me two glasses.'

'Hi handsome,' said Dorothy. Her eyes appeared glassy. 'What's de latest?'

'Gyul, meh life too exciting. Yesterday was meh luckiest day in meh life. Imagine

de duck lay ah set ah eggs and later dat day, a car kill de duck, poor ting. Is a good ting I get the eggs before dat. Ah put de dead duck in de freezer so on Boxing Day ah cooking curry duck and dhalpuri.' He licked his lips as he thought of the cooked meal.

Marcus added pepper and salt to the oysters and gave the glass to Bally. 'De devil does work in mysterious ways. Doh forgeh to invite meh when yuh cook it.' He looked at the passing cars and hoped for more customers.

Bally raised his hand. 'We go see. It is a small duck.' He handed the empty glass to Marcus. 'Another one.'

Marcus nodded. 'Meh mind tell meh so.' He placed some oysters in the glass and began to add pepper. 'Yuh hear Philip dead?'

'Yeah ah saw it in the death announcements,' replied Bally. 'He was a good man.' He scratched his leg and complained about the mosquitoes.

Marcus smiled. 'Yes anytime somebody dies people would say she was a good woman or a nice man. It look like nobody bad does die.'

Dorothy laughed. She asked, 'So Mr. Bally what are your plans for dis festive season?'

Bally looked at her short skirt and touched a scar on his forehead. 'Well so far de year treating meh nice. Since July ah was able set off four small bombs in de city and de police eh ketch meh as yet. Ah hoping to set off a bomb next week. Close to Christmas, so dat people will be thankful to be alive and go to church, and remember de birth of Christ. And give thanks to God they were not targeted.'

'Just amazing. Doh worry de secret safe with we. Remember we is we and after we is weevil. You are doing a splendid job creating kutchoor. People are becoming more religious because of all these bombings. Buh keep in mind- dat closer to church further from God.' Dorothy was a gullible person.

Bally enjoyed the compliments and continued to exaggerate the impact of his actions. 'Yes, and fathers spending less time liming at rumshops and other places and are now spending quality time with their wives and children at home. Dese days dem worse dan Mr. Cyril. Remember is because of my bombings dat people are able to forget about the high number of murders and kidnappings. It keeps their mind at peace.'

'Bally is people like you dat deserve a national medal or statue. You jus' like dis government with their smokescreens and red herrings trying to distract citizens from real issues. You trying to keep de family together and bring religion back into the lives of our citizens and create a peaceful place,' said Marcus. He bit into a ripe five fingers. It was his favourite fruit.

He did not detect the cynicism. 'Is ah very difficult job but somebody must do it. There will be some collateral damage.' He looked at the ground and appeared in a pensive mood. 'Ah is ah very humble person and hate publicity so not many people know dat I am responsible.'

Marcus knew that Bally was lying. He was aware that Bally had a reputation as a coward and a pathological liar. He sarcastically commented, 'Ah sure dem bubulups women will go jus' crazy over you if dey realize dat you risking yuh life to save dis beautiful, rainbow country.'

Bally nodded. 'Yes, dey well love me. If I tell de public ah would be more popular dan de Strike Squad, Soca Warriors and dem athletes who win Olympic medals. I am also assisting tourism. Because of de bombings, ah put Trinidad on de world map. Now everywhere know about meh bombs. Nex' month some tourists coming from Scotland Yard, CIA and de FBI to check Trinidad out. Ah feel dese people from de FBI might want to put up hotels or open a casino or something.'

Dorothy shook her head. 'Is true dat de devil does really find work for idle minds and hands. Did you feel jealous in October when de police and odder people blame gangs for de bombings?' She wanted to leave for her home to eat fruits. She had two bowls of chennet and cerise in the refrigerator.

'A little, but everybody trying to be a hero like me. So much good for odders to do. Not many people know dat some members of de Special Branch of our Police Force knew about the planned coup in July 1990 but dey did nothing. Anyway, de prime minister listen to my recent advice and had a meeting with gang leaders at Crowne Plaza Hotel. He give them food and drinks paid for by taxpayers and make dem sign an agreement. Of course we know dat is ah public relations gimmick. Crime will continue.'

Marcus was tired of his boasting. 'Dat is fantastic. No wonder nobody has convicted you as yet! Have you ever worked with international security agencies?'

Bally again did not detect the sarcasm. He felt Marcus genuinely admired his work. 'Yes, and also with de Mossad from Israel. Remember the expensive crime unit patrolling in de sky and de Blimp?'

Marcus nodded and avoided eye contact with Dorothy and Bally.

'Well, it was me who arrange for de government to buy dem tings. And leh meh tell yuh a lil secret- de expensive Blimp dat flying all over de country don't have any equipment to see crime. It empty! I is Mr. Big who de prime minister talking 'bout and I is a big fish in the criminal underworld. Nobody could touch me because I have files on some key government ministers. De police real lame because they need breakfasses. To help fight crime yuh need healthy police. Yuh agree Dorothy?'

She nodded. 'Yes, yuh sounding like a future Minister of National Security. Jus' amazing. Is people like you dat making sure dis country achieve 20-20 vision and developed world status like Canada and United States.'

Bally thumped his chest. 'When I in charge, who doh hear go feel. Puh dat in yuh pipe and smoke it! So sweetness you know what I want Santa Claus bring for me dis Christmas?'

'No. But —ah feel yuh want ah piece of pork for Chrissmass.' Dorothy smiled and touched his shoulder. 'Yuh sure dat is all yuh want? I have a clean neck fowl waiting to cook.'

Bally's eyes were wide open. 'Well, ah like what Scrunter also say –ah want meh breakfass in meh bed.' He grinned and traced the length of the scar with his index finger.

Dorothy had a mischievous smile and placed her arm around his waist. 'Yuh mean breakfasses.' Breakfasses was a word mistakenly used by the Minister of Education when she commented on the plural of breakfast.

'Yeah dat too.' Bally laughed and this exposed two black molars and three gold teeth in his mouth. He placed ten dollars on the table. 'Son, dis for de oysters.' He watched the sky and then stared into Dorothy's eyes. 'Such a beautiful night with de stars in de sky.' He was going to say something romantic then remembered Marcus was listening. 'It really look like it going to rain plenty tonight, well me doh want to get wet and you ketch cold, so is real nice talking to all yuh. Anyway ah going now ah have to feed de dogs. Dorothy doh forgeh to come tomorrow night for parang practice by meh house.' He winked at her. 'I is ah chac chac player.'

'Okay sweetheart. Ah leaving also.'

Marcus wiped the table with a rag. He glanced at the two departing figures. He did not like Bally. Residents often said that Bally used the excuse of parang practice to seduce impressionable women. Two years ago Dorothy's husband discovered that his wife was unfaithful. He accused Bally and attacked him with a cutlass. Bally narrowly escaped but received a chop to his forehead.

On Sunday evening Marcus played football with his friends. Nigel had built two goalposts with plastic pipes and old fishing nets. These goalposts were used for small goal football on the street and after the games were stored in Anil's garage. The games would be three-a-side or four-a-side. Whenever there were more than six persons the game was moved to the nearby savannah. The weekend games usually attracted persons from nearby Longdenville, Endeavour and Cunupia. Their games usually comprised fourteen or sixteen young men. This would lead to teams of seven or eight persons playing against each other for ten minutes. At the end of the stipulated time, the side with the most goals would remain and the losing side replaced by another team of seven persons. Sometimes the players agreed that the victorious team would be the one that scored the first goal.

Pushing a player, a rough tackle and pulling a rival's jersey often lead to arguments as players sought to determine if the incident constituted a foul. The announcement of 'last play' by a timekeeper would increase the urgency of scoring a goal. And, after the ball crossed the boundary of play there would be penalty kicks.

One evening, Marcus returned home and went to the living room. He began reading an application form for an undergraduate degree at the University of the West Indies. He had chosen Management of Business and Accounting. He stopped reading and stared at a black and white photograph of his grandfather. He heard stories that his grandfather was one of the country's best taxi drivers. Marcus wished his grandfather was alive to advise him on a career.

Chapter 5
Uncertain roads

All the boys, except Marcus, moved out of Enterprise and played for local football teams. They would sometimes return to Enterprise to celebrate birthdays or to spend Friday evenings relaxing in a rumshop. Marcus was chosen as a midfielder by the coach of Lendore United. After six months he quit the team and joined Crown Trace B. He played against Todd's Road United, Impact FC, Double Dog, Boothill All Stars, Cool It and Innocent Fun. These were teams in the Lendore Football League and the games were played at the Chrissie Terrace Recreation Ground.

Quasee played as a defender for Joe Public in the National Super League. In 2001 he was named 'Most Valuable Player' and received a trophy, medal and five thousand dollars. Quasee had faced teams such as Crab Connection, Rangers and WASA. His most memorable year was 2002 when he scored a beaver-trick against Club Sando. In 2004 he stopped playing football due to an injury he received in a car accident.

Anil was a midfielder for Jaboleth in the T&T Professional Football League. He played against other teams as Defence Force, W Connection and Starworld Strikers. Aboud was the reserve goalkeeper for Superstar Rangers in the National Super Football League. He saved goals from strikers of Paragon, Malvern, La Famille, Maple, and Fire Services. Ming was recruited as a striker for Petrotrin in the Southern Football Association. He earned the nickname of Pele because he regularly scored goals against Barrackpore United, Giants, X-Men, Penal All Stars and Palo Seco. In 2001, Ming married and moved to Santa Cruz. For two years he was a defender for Hot Shots that was included in the Santa Cruz Football Organization Monteco Cup. The games were held at the Brian Lara Recreation Ground at Sam Boucaud. Ming played against teams such as Pipiol, Back Street, North Coast and Bourg Mulatresse.

Gabriel was employed as an assistant coach for Alcons in the Eastern Football Association. He tried to be on the team but the other players realised he never wanted to pass the ball. The footballers realized that Gabriel was only interested in receiving praises. He did not appreciate the value of the team's effort. He did not want to share any fame. Alcons would face teams as Young Hearts, HCL Strikers, BM Spurs and TSTT.

Kervin played as defender for Las Lomas Youths. He was also captain of the team and defended his goal from players of Juventus, Couva Players United and Leeds in the Central Football Association. He eventually moved to Tobago to work at a hotel. He began playing as a midfielder for Bethel United and met teams including Pepsi Hills United, Sidey's, Mason Hall PYC and HV Milan. During his four years at Bethel United he scored two goals. His younger brother, Bobby, played as a midfielder for Charlotteville Methodist in the Milo Tobago Primary School Football League. Bobby scored goals against Ebenezer Methodist, Belle Garden Anglican and Delaford Anglican.

Aboud, Anil and Marcus were keen supporters of wrestling. In 2004 they went to the Centre of Excellence in Tunapuna to view a wrestling match. The crowd eagerly waited for Thunderbolt Williams, the local wrestling champion, to fight The Invader of Puerto Rico. Thunderbolt won after eight minutes. The older members of the audience were eager to view Abdullah 'The Butcher' of Sudan clash with Victor Jovica of Croatia. Both wrestlers were disqualified for fighting outside the ring. In another match, Carlitos Colon lost to El Bronco after six minutes of fighting. Other wrestlers who fought in the event were Savage, Exterminator, Strongbone and Hercules.

Aboud was angry that the main fight between Jovica and Abdullah was short. He demanded a refund. The owner refused and Aboud struck him. The police were called and he spent the night in jail in the Diego Martin Police Station. Two months later Ming attended the Caribbean Football Union Championship at the Hasely Crawford Stadium. Ming was more interested in local rather than international football. It was a first-round qualifier with Caledonia AIA, a Trinidad and Tobago Digicel Pro League club, playing against Newton United, a St. Kitts-Nevis club. Newton United launched a series of quick counterattacks but could not penetrate the Caledonia defence which was protected by Nuru Muhammad and Daniel Cyrus.

Keyon Edwards of Caledonia was on the edge of the box when he fired a left-footed shot. It floated over the wall of defenders but was brilliantly stopped by a two-fisted save from goalkeeper Akeil Byron. The Trinidad and Tobago club eventually won the game by one goal. This goal was a penalty taken ten minutes before the final whistle.

In 2005 there were six teams in the CONCACAF group that were hoping to qualify for the 2006 World Cup in Germany. The teams were Costa Rica, Panama, Guatemala, United States, Mexico and Trinidad and Tobago. On 9 February, at the Queen's Park Oval, Trinidad and Tobago would meet United States. Marcus, Aboud, Anil and Ming were among the fans. Anil had won two tickets for the Carib Stand. Marcus and his sister were in the Republic Bank Stand and Ming was in the Trini Posse Stand. Aboud was comfortably seated in the RBTT Bank Stand.

Anil looked at the sea of red supporters at the Oval. There was a feeling of expectation. Along the edges of the field were billboards advertising Cable and Wireless, Suzuki, Solo, Sissons, Shell, Royal Stag, FedEx and Pepsi. There were some empty chairs in the Learie Constantine Stand and C.L. Duprey Stand. During international cricket games he would usually sit in the Errol Dos Santos Stand and view the WITCO scoreboard after every over. From that vantage point he was able to see the top of Queen's Royal College and the red sign of Tatil Insurance.

The score was 1-0 in favour of United States. The crowd was disappointed but continued cheering the Trinidad and Tobago players who were dubbed the 'Soca Warriors'.

Ming was disappointed that the Soca Warriors did not score the first goal. He left the stand and bought a snocone. 'Put some condensed milk in it.' He placed five dollars on the vendor's cart.

The vendor nodded and took a can of condensed milk and allowed it to slowly drip on the snocone. He inserted a straw in the snocone and gave it to his customer.

Ming slowly walked up the stairs and glanced at the Gerry Gomez Media Centre. He remembered that he had to apply for the advertised jobs as a journalist at the *Express* and *Newsday* newspapers. He needed to update his resume.

Aboud was waving a mini Trinidad and Tobago flag. He felt dizzy and wondered if his glucose level was low. The aroma of food made him hungry. He left the stand and bought a bake and shark sandwich and a cup of corn soup. He returned to his seat and began eating.

Brian McBride of the United States jumped high to head the ball towards the goal. Two of the Soca Warriors, Marvin Andrews and Kenwyne Jones, stood nearby and appeared shocked. Marcus watched in horror as the ball slipped past the outstretched hands of the goalkeeper and entered the goal. The Soca Warriors seemed to be moving in slow motion as the speeding ball was stopped at the back of the net. This would not be the first time that the crowd would be silenced. The final score was 2-1, the United States had defeated Trinidad and Tobago. Once again the expectation and joy of a wedding had become the sadness of a funeral.

The three teams with the highest points would qualify for the World Cup finals in Germany. Mexico and United States had earned sufficient points to qualify. The Trinidad and Tobago football team had to defeat Bahrain in the Asian/CONCACAF playoff. Two games were to be played- one in Trinidad and Tobago and the other in Bahrain. The winning team would qualify for the World Cup.

The Soca Warriors, included Dennis Lawrence, Marvin Andrews, Avery John, Stern John, Dwight Yorke, Silvio Spann, Carlos Edwards, Russell Latapy, Kenwyne Jones and the goalkeeper was Kelvin Jack. For some fans, the odd man was the midfielder- Christopher Birchall. He was the first White player to be on Trinidad and Tobago's team for the past sixty years. Birchall played for the Port Vale football team.

The first game against Bahrain was held at the Hasely Crawford National Stadium in Trinidad. Despite previous disappointments, the vast multitude of faithful fans never abandoned their fickle heroes. Scalpers outside the stadium were selling tickets for four hundred dollars. Bahrain's Ghulum Ali headed the ball past Kelvin Jack into the net. The crowd became nervous and anxious. Ten minutes later there was a long range shot at the Trinidad and Tobago goal. Kelvin quickly scooped the ball off the ground and held it close to his chest. He glared at an approaching Bahrain player. Then in the seventy-sixth minute, midfielder Chris Birchall, who was twenty metres from the goal, blasted a shot that rocketed into Bahrain's net. The sell-out crowd of 30,000 diehard fans erupted. Their pent-up emotions exploded. The final score of 1-1 kept alive the hopes of both teams.

The second game in Bahrain would determine the fate of both teams. The Laventille Rhythm Section and Woodbrook Playboyz travelled to Bahrain. They planned to provide music for the Soca Warriors in the stadium. However, due to security concerns, the Bahrain Football Association banned both musical groups from entering the stadium. It seemed that without music the Soca Warriors, with the white jerseys and black pants, would be powerless. Music was the sustenance of the fans and footballers. It heightened their emotions and increased the excitement.

In the forty ninth minute, a corner kick was taken by Dwight. Dennis Lawrence saw the ball in the air and jumped towards it. He headed the ball into Bahrain's goal. The ball at the back of the net meant a giant leap forward for one small Caribbean nation. Bahrain stood still. That goal would be the final nail in their coffin. The final result was 1-0. On this occasion, scoring a goal produced an instant hero and ensured a place in football history. The ball at the back of the net had given sainthood to the entire football team. In football, being at the right place at the right time plus a little luck was the recipe for an instant hero who would soon be forgotten. There was no need for long hours of study, patience of a chess player or discipline of a bodybuilder.

The victory over Bahrain produced jubilation in San Fernando. The daytime celebrations included glaring headlights, blaring horns and exchanging of high fives. At Black Gold's Bar on Coffee Street, fans began chanting

'Germany, Germany, Germany.' Enthusiastic fans in Knockers Bar, on Sutton Street, ran outside to spread the good news. They waved flags and shouted 'We win! We win!'

The qualification of Trinidad and Tobago for the World Cup in Germany meant that Trinidadians and Tobagonians would be making the trip to Germany to support the Soca Warriors. This was a pilgrimage to the holy land of football.

The returning football team assembled in the VIP Lounge of the Piarco Airport. They met local football officials and media personnel. Outside there was much fanfare with a large sea of red. Supporters had donned red caps, jerseys, pants and dresses. Nearby were trucks with soca music blaring from the music speakers. The unofficial anthem was 'Soca Warrior' by Maximus Dan. It was a Carnival atmosphere with moko jumbies and masqueraders from Poison. The crowd wanted to touch their heroes. Some members of the crowd had outstretched hands and pushed to be closer to their favourite footballer. Some in the crowd wished this moment would never end.

It was an important day for Marcus. Raphael would be released from prison. Marcus decided he would carry his father and Anil to the airport to welcome the football team. Raphael wanted to go Germany to view the games and support Trinidad and Tobago. However, he did not have a passport.

Anil touched Marcus to get his attention and then pointed to Jessika.

She wore a small red jersey, tight red jeans, gold chains and over-sized earrings. She was screaming the name of Dennis Lawrence.

'Who is dat?' asked Raphael. 'Ah want to struggle with dat structure.'

Anil smiled. 'Yuh remember little Jessika from Enterprise?'

Raphael looked surprised. 'Dat is Jessika? She looking so force ripe.' He paused. 'Not me and dat…or else ah go make ah jail again!'

Marcus nodded. 'Ah expected she to be here. Some of dem people acting as if their problems go disappear if they touch these footballers. It seemed as if the sick would be healed if they touched the footballers.'

Dennis Lawrence, the hero, spoke. He told the crowd that the crime wave must end in Trinidad and Tobago, 'This is the start of what can only be peace, love and harmony in Trinidad and Tobago.' It was a short statement and for many persons it was a brilliant speech.

Government officials who were previously unwilling to assist were now eager to provide financial assistance. These hypocritical politicians merely sought to improve their public image. They did not truly care about the development and progress of football in Trinidad and Tobago.

'Dem Soca Warriors is de messiahs. Dey will soon take us to the promised land of the World Cup,' said Anil. 'Boy ah could imagine how much girlfriends dem footballers will now have!'

Raphael remained quiet and surveyed the animated crowd. He did not want to speculate on the future of the football team. He hoped they would give a sterling performance at the games in Germany. He hoped that the unity generated by the football team would continue.

The motorcade with the football team departed for Port-of-Spain. The team was the pied piper as thousands of excited fans obediently followed them to Port-of-Spain. These fans would have followed their heroes around the world. For five hours, Quasee, Bally and Dolly endured traffic. Thousands of fans passed one of the advertisements of the Nestle factory located in Valsayn. It was an ad of a man kicking a football.

Bally, Dolly and Quasee were soon among thousands of persons at the Brian Lara Promenade. They were patiently waiting for the grand motorcade. Time and lateness had no meaning at the moment. There was live entertainment provided by popular artistes. At 10 pm the motorcade entered Port-of-Spain. A fan waved a hand-painted sign which had the crooked words '2006 World Cup finals Trinidad vs Brazil.'

Dolly told Bally and Quasee, 'Tomorrow is 19 November. All yuh remember why dat day important in we history?'

Both men nodded.

Bally smiled. 'It was a day when big men cry. It was when the United States beat we by one goal. I was there. It was a day when no ball should have been in the back of our net.' The smile slowly disappeared and he became pensive.

Dolly replied, 'Well today is Redemption Day.'

Bally, Quasee and Dolly heard the familiar chant- 'Ole, ole, ole, ole, ole. I am a Soca Warrior.' Then they saw the brightly decorated truck with the Soca Warriors. The crowd began cheering. The local dignitaries and the footballers slowly assembled on a temporary stage.

Carib Beer, the local brewery, organized the 'I Going Germany' competition in which one hundred persons would be given all-inclusive packages of spending money, airfare, accommodation and most importantly- tickets for all the Soca Warriors games.

Quasee and Anil decided they would try their best to be among the 100 winners going to Germany. Every week Quasee purchased ten cases of beers and each day would drink seven bottles. Quasee's father and two brothers assisted him in drinking the beer. Every week Anil purchased three cases of beer. He would say a brief prayer before opening each bottle. After two months he began to complain of blurry vision. He had diabetes and was not aware of the symptoms.

Ming visited Scotiabank and applied for a loan. He told the bank officer he wanted the money to repair his home. He wanted a loan of twenty thousand dollars. This would be sufficient to pay for the trip, hotels and entrance fees to the games. However, due to a lack of collateral his loan application was rejected.

Aboud wanted to buy all the memorabilia of the Soca Warriors. He purchased trolley bags and knapsacks which had Russell Latapy's signature. These items were in the national colours- red, black and white. He paid two hundred dollars for collector's coins produced by the Central Bank. These silver and gold coins were too expensive for the average fan.

Gabriel, Kervin and Aboud decided they would buy Lotto tickets every Friday. Each person would buy four tickets and promised that if he had the winning numbers, he would use the money to pay airfare and accommodation for the other two friends.

One evening after playing football, Kervin told Aboud, 'We cannot just blindly buy Lotto tickets. Boy we have to seriously start choosing numbers that would win. We cyah buss ass.'

Aboud looked confused. 'How we go do that? Should we go by de obeah man from Moruga or de psychic from Mexico?'

Gabriel was eating peewah. 'Not dat Mexican, she false like meh grandmudder false teeth. All she predictions is ah waste of time. Tomorrow we will visit de obeah man.'

Kervin was serious. 'No. Nothing about obeah man and psychic woman. We need something more powerful dan obeah and seer woman. See what numbers associated with us. Take for instance, this evening you score two goals and ah score five. So these two numbers could be used for the tickets on Friday.'

Aboud shook his head. 'Yes, and we could also use the numbers on the jerseys of the football team. Lissen fellas ah thinking 'bout becoming a wrestler after dis World Cup.'

Kervin was excited. 'Good yuh now understand. We could also use the numbers of your house, telephone number, passport number and licence number of your fadda car.'

Gabriel was deep in thought and oblivious to the conversation between Kervin and Aboud. He was deciding on his lucky numbers. After fifteen minutes, the young men departed for their homes. Each now felt he had a better chance of winning.

That evening, Aboud began making a list of numbers which were associated with his daily routines. He opened the fridge. There was a sandwich loaf with fifteen slices. He wrote fifteen on a piece of paper. He ate a grapefruit and it had nine seeds. He looked around the kitchen. There were four apples and two bananas on the counter. He shook his head and noted the numbers.

Marcus felt he would not win a trip to Germany and did not have sufficient money in his bank account to pay for airfare and accommodation. He wished his mother's flower shop was more profitable. He decided he would watch the games on television.

Gabriel wanted to be in Germany for the World Cup. He decided to sell one of his kidneys. He sold his left kidney to an American citizen for eighteen thousand dollars. This would be enough money for him to go to Germany and support Trinidad and Tobago in their three games in Group B.

Ming became desperate and decided to commit robberies to finance his trip. He met a car salesman from Bamboo who offered him five thousand dollars to steal two cars. He decided to accept the offer. After receiving the payment, Ming sought other illegal means to earn money. He decided to rob Carib Brewery trucks. Within two weeks he had stolen seven thousand dollars from four trucks.

The Carib Brewery Company decided to paint large signs on their trucks- 'No Cash on Board. We are now Cashless.' Ming was not bothered. He had sufficient money for his trip to Germany. However, thieves who wanted money and were illiterate, continued to stop and rob the Carib Brewery trucks.

Anil checked under the cover and saw the word- 'Winner.' He felt weak. He kissed the cover and began to jump up and down. His prayers had been answered and he rushed upstairs to retrieve two suitcases from a cupboard. They were dusty. He carefully carried both downstairs and placed them in the sun. Later that evening he wiped the interior of each suitcase. Anil nervously checked his passport. If he needed a new one, it might mean a delay and a cancellation of the trip to Germany. It would expire in 2012. He kissed the page and telephoned Marcus.

Chapter 6
Victory and sad realities

Anil, Gabriel and Ming were staying at the Dortmund Hilton in Germany. Anil and Ming were in the lobby waiting for Gabriel. They met an elderly football enthusiast, Nwankwo Baffoe, from Nigeria. Nwankwo shook their hands.

Nwankwo initiated the conversation. 'So where are you from?'

'Trinidad and Tobago…in the West Indies,' said Ming.

'Is your country in the World Cup or are you like myself, coming to see all the games?' asked Nwankwo.

Anil quickly responded. 'Our country made it to the World Cup finals for de first time. We cannot afford to see all the games. Is your country represented?'

Nwankwo paused. 'No, Nigeria did not qualify. I have also lived in Ghana. It is the dream of every little Black boy in Africa to play professional football. They want to be the next Didier Drogba or Michael Essien.'

'Yes I've seen Drogba and Essien play,' said Ming. 'Do you know about de legendary Eusébio da Silva Ferriera of Portugal?'

Nwankwo's eyes lit up. 'Yes, the Black Pearl! Have you seen or heard of the World Cup quarterfinal in 1966?'

'I read about it.'

'I saw it. It was a fast-paced game and North Korea was leading 3-0. Then Eusébio slammed four goals into Korea's net. But let me continue about football in Africa today. In Ghana there are many unlicensed football academies that charge high fees to train boys full-time. Some of these academies even operate by the roadside.'

'How do parents afford this?' asked Ming.

Nwankwo glanced at one of the hotel's employees watering a plant in the lobby. Yesterday he wondered if the plant was fake or real. He looked at Anil and Ming. 'Some parents sell their homes and move to the city to enrol their children sometimes as young as seven and eight years in these academies run by conmen. Sometimes parents give their life-savings for the trip to France, Spain or England so their son could go on trial for a club.'

Anil looked surprised. 'Is so for odder states in Africa?'

Nwankwo nodded. 'Well a few years ago the captain of Ghana's national team, Stephen Appiah, was a midfielder for a club in Turkey. He is now a millionaire and his family is happy. For many Blacks in Africa, having a son playing professional football is like winning the lottery. Football is not simply kicking a ball. The game is the great equalizer. When Mandela and others were in Robben Island prison, he and other inmates demanded they be allowed to play football. For three long years they asked and were denied until the prison agreed to allow thirty minutes on Saturdays.'

Gabriel entered the lobby and greeted Nwankwo.

Anil rose from the couch. 'Good, let's get the bus to the stadium.'

'Good luck to your country in the World Cup,' said Nwankwo.

Football is similar to the art of war. The football field is transformed into a battleground where a developing country has an opportunity to defeat a strong developed nation. The ball on the field is a cannonball which could be used to destroy a country's nationalism and pride. The players are serious as soldiers and armed with skill and vitality. They have been trained to return victorious at the end of their mission.

In Germany, the official bus for the Trinidad and Tobago team had the slogan 'Here come the Soca Warriors- the fighting spirit of the Caribbean.' In ten minutes Trinidad and Tobago's Soca Warriors would play Sweden in Group B of the World Cup match in Dortmund, Germany. For many citizens of Trinidad and Tobago, June 2006 was an important milestone in the country's history.

Marcus and Kervin were at the upscale Trotter's Restaurant on Maraval Road in Port-of-Spain. Each patron paid six hundred dollars to view the games and was given food and drinks. Both men were not concerned with the cost and wanted to view the game on a large screen. About four hundred persons, dressed in one or more of the national colours, were at the venue. Some businesses such as Smokey and Bunty, Crobar and Sweetlime had televisions outside for some of their patrons and the public to view the games.

Kervin quietly steupsed. 'Dis place so damn noisy. How we go enjoy de game?'

Marcus was disturbed and shook his head. 'Yes and dis blasted cigarette smoke. Leh we jus' focus on de game and forget about dem drunkards.' He loudly blew his nose in a blue handkerchief. 'Lissen, yuh tink we could draw or win dis game?'

'Ah not too sure. De Soca Warriors didn't look too good in dem warm-up marches against Slovenia and Czech Republic recently. How yuh dad enjoying being a free man again?'

Marcus rubbed his eyes. 'Yeah meh fadda top ranking. He want to live by himself, he always like to be a sufferer. Today he gone to a funeral...a funeral for the amputated foot of his Uncle George. Dey even buy a small casket for de foot. Ah tell yuh dat whole family weird.' He looked at the other patrons who were all excited. 'Ah hope de Soca Warriors cause ah big upset. Yuh hear dat next year David Beckham leaving England and going to play for Los Angeles Galaxy in the United States?'

Kervin sipped his drink. He did not enjoy discussing funerals. 'Yeah ah read something about him going to play in the Major League Soccer. Yuh read de prime minister's statement dat crime is temporary?'

'Yeah. He and the damn government will be temporary if he keep on making dotish statements. Imagine de Minister of National Security say dat kidnappings on de decline and murders increase because kidnappers change their profession to murderers!' Marcus took a handful of salted channa from a bowl on the table. 'All ah dem politicians talking garbage. If politicians were being killed or kidnapped yuh would see how quick dis crime problem would be under control. Notice how during Carnival de police and army does have crime under control but right after Carnival de crime does continue.'

The song 'Shaka Shaka' resonated throughout the restaurant. It was played anytime the goalkeeper, Shaka Hislop saved a goal or held the ball. There was shouting among the fans when striker Stern John had the ball and was advancing towards the Swedish goalkeeper. He was tackled and lost the ball.

Kervin became worried when Avery John was sent off the field for a second foul against Christian Wilhemsson.

'How we go win with ten men?'

Marcus was unperturbed. 'Think positive. Avery is the first player to be sent off in dis World Cup final. We in the record books again!' He was satisfied with the ball possession of the Trinidad and Tobago players but concerned that Densil Theobald was slipping on the field.

Cornell Glen was introduced to replace Collin Samuel. Cornell on his first touch of the ball decided to back-heel it to Dwight. Five minutes later Cornell almost scored a goal when he ran into a headed-on pass from Stern John. Dwight took a corner kick and the ball curled towards the Swedish goal. A stocky Swedish defender jumped and headed the ball safely away from the goal. A disappointed Dwight wiped his brow and adjusted his shin pads. Four minutes later, he fired a shot than went over the top bar.

At the Dortmund Stadium, in Germany, there were four thousand fans from the West Indies. There was music from Machel Montano and Xtatik. Chris Garcia was singing his song- Deutchland. The result was a goalless draw. This would seem to be normal score for a football game. But this was the World Cup. More importantly, the score was achieved by a team which had never played at this international level. Trinidad and Tobago had been described by other teams as the underdogs and the team of old men.

Anil was elated. The mighty minnows played most of the second-half with ten men to earn their first ever point in a World Cup final. The 67,000 persons who witnessed the game at Dortmund Stadium would have rarely seen such courage and fighting spirit.

Ming told Gabriel, 'Dwight deserve to be Man of the Match for being a playmaker. He real work in de midfield but the real hero was Shaka Hislop who guarded dat entrance of de goal.'

'De whole team deserve de award. Everybody play good to defend dat scoreline especially de defenders like Cyd Gray,' said Anil. 'Remember Sweden has been 11-time World Cup qualifiers. Is we furse time.'

Gabriel lit a cigarette. 'Yeah dat is good talk. Ah sure odder people think de same way 'bout de defenders. Buh Hislop was at his best today. Yuh wants nothing with him.'

'All yuh going to sample some German beers?' asked Anil. He coughed. The weather was cold. 'Ah wonder how much Euros a beer cost?'

Gabriel grinned. 'Must be two or three Euros. It shouldn't be too much.' He looked at the departing fans. 'Fellas ah want to go and check out some of de German girls.'

Anil laughed and patted Gabriel on his back. 'Aye boy behave yuhself… anywhere in de world yuh is a real sweet man! Yuh have long eye.'

On the morning of 15 June, Anil awoke at 4am. The room was cold. He switched off the air conditioner but could not sleep. He decided to read the football websites on the internet. Many fans believed 15 June was another landmark in the history of Trinidad and Tobago. The team would face mighty England. He was surprised to see that the odds for Trinidad and Tobago to win were 28 to 1 and a draw was 20 to 1. He was not surprised to read that England's captain David Beckham saw Dwight as the danger man in the clash between the two countries.

Dwight had made a sacrifice by coming out of retirement. British fans and the media had given him the nickname 'Smiling Assassin' due to his deadly accuracy in shooting at the goal and his broad smile after scoring a goal. He was largely responsible for Manchester United winning the UEFA Champions League, English Premiership League and FA Cup.

In many countries, football is one of the most important factors defining masculinity. Boys and teenagers aspiring to be famous often seek the avenue of football to earn the admiration of peers and the approval of society. More importantly, for fans, their aspirations and knowledge of the game comprise a special domain promoting bonding. This group identification with a male-dominated game is an important aspect of masculinity. During football, the divisive differences of class, culture, nationality, gender, ethnicity and religion are temporarily set aside.

Anil, Gabriel and Ming arrived early at the stadium in Nuremberg. They wanted to avoid the long lines. Ming rubbed his hands together and shivered. Both Anil and Gabriel wore thick sweaters and gloves and did not seem bothered by the cold weather.

'Yuh think we could draw this game?' asked Ming.

Anil shook his head. 'Draw? We could win because Michael Owen didn't score in the Paraguay game. Like he out of form. Ah feel we can cause an upset in de game. Yorke say dat we not just here for de ride we going to beat England and make history.'

Gabriel was pensive. 'Yeah dat big upset which we want could happen and it seems Wayne Rooney out of the picture. Ah hear he injured.'

The three friends entered the stadium and took fifteen minutes to locate their seats. Ming said, 'Yuh hear dat Angostura offered twenty-four barrels of rum to de entire team if dey beat England today.'

'Ah not surprised. Dem companies want all ah we to be alcoholics,' replied Gabriel. He stared at a lady waving a German flag. 'Before ah leave dis place ah have to geh some white meat.'

Anil pointed to one of the spectators. 'Dat fella in de red sweater looking like John Barnes.' He adjusted his spectacles.

Ming stared at the spectator. 'Nah. Barnes would never be sitting in these rows. He would be near the VIP section or serving as a commentator. Dat man/ was one of the best left-side midfielders who ever play for England. He was a legend when he played for Liverpool and did help dem win de FA Cup and League Championships.'

Anil replied, 'He fadda from Belmont. In fact wot ah hear was dat a few years ago Barnes was shortlisted to be de coach for de Soca Warriors and we Football Federation decide to give it to de Columbian coach- Maturana. Imagine Jamaica and Nigeria turn him down because of a lack of experience.'

'Belmont? Yuh sure Barnes have Trinidad roots?' asked Gabriel. He scratched under his arm. He was allergic to the deodorant. 'Every famous person in dis world does have some connection with de Caribbean either they visit or born there.'

Anil nodded. 'Yeah he from Trinidad. He face a lot of racism when he was a footballer and even when he retire it was difficult for him to geh a coaching job.'

In Trinidad and Tobago there was a high level of excitement. One day before the grand showdown between England and Trinidad and Tobago, representatives from the Ministry of Sport distributed free caps, T-shirts and bandannas in Montrose, Chaguanas. Marcus, Cleopatra, Harriet and Malcolm were in the crowd. They received caps and bandannas. Persons had informed their employers they would not be at work on 15 June. Some employers decided to close early and depart to their homes to witness the game.

Marcus glumly stared at the television screen. He wanted to be amongst the red-clad Trinidad and Tobago fans in the stadium. As the game progressed, he could not believe his eyes. Eighty two minutes had passed and the score was goalless. Dwight fired a shot but the ball was not on target and hit the right side of the net. The nightmare began when England's striker, Peter Crouch, grabbed the dreadlocks of Brent Sancho and climbed on his back to head the ball past Hislop. The Caribbean-born goalkeeper regained his composure. Caribbean fans and some other supporters were stunned and expected the referee to stop the game and issue a yellow card and disallow the goal. Marcus loudly shouted expletives at the referee and linesmen. The ball did not seem to want to stay at the back of the net. The ball wanted to leave the net. The ball seemed guilty as if it had committed a wrong. It wanted to be free again on the field.

This goal shattered the dream of one man and the dreams of thousands of fans in the Caribbean. The first shot had hit the target. The Smiling Assassin was powerless. The goal made some believe football was corrupt. The ball at the back of the net tested the unity of a tiny Caribbean nation. The goal wounded the West Indian ego and broke the resilience of the Soca Warriors. Their fate was sealed with a second goal coming from midfielder Steven Gerrard.

Anil was infuriated. 'Is no wonder dat United States had to bomb China… Japan…Korea…wherever de hell de referee from. Dem kinda people dotish. Dat referee is blind as a bat and a big coward. He 'fraid to rule against England. He should only be allowed to referee matches played by school chirren.'

Ming shook his head in dismay. 'Well dat is it for de Soca Warriors. Ah hope somebody from FIFA complain 'bout de bad refereeing. Somebody should shoot de referee. Dem kinda ting does psychologically affect we players.'

In the third match, at the Fritz-Walter Stadium in Kaiserslautern, Trinidad and Tobago was beaten 2-0 by Paraguay. The day in the sun for the Caribbean nation had already gone. There were moments when the Trinidad and Tobago fans expected a miracle goal from their gods. In the fifth minute Dwight took a free kick and Cornell jumped to head the ball in the goal. However, the Paraguayan goalkeeper punched it away. Later in the game the unthinkable occurred. The ball glanced off Brent Sancho's head and rolled into the goal of Trinidad and Tobago. This was one of the occasions when scoring a goal did not produce an instant hero. However, it was another record for the small Caribbean nation. It was a humiliating defeat for a team that had given mighty England a scare and held a strong Swedish team to a draw.

Gabriel looked at his watch. He was concerned that Trinidad and Tobago did not have sufficient ball possession in the second half of the game. He realised that whoever possessed the ball had the eyes of the world. The dreammakers would be the team scoring the goals. Three more minutes for the final whistle and then calypso football would end its glorious dance on the world stage.

Anil was staring through his binoculars. He pointed to a section of the stadium. 'Look is Justin, Pandey, Ramsaran, Richards and Boines sitting down there.'

Ming remained transfixed on the game. 'Yeah dat is for VIP big pappy not like me and you.'

The referee's whistle signalled an end to the World Cup dreams of Trinidad and Tobago. Most of the fans of the losing team would return home.

Ming shook his head. 'Ah disappointed with de coach. Ah feel he should have played Latas more. The Little Magician should have played a lot in all three games. We could have drawn dat game with Paraguay. Look how he explode in de sixty fifth minute. Look at de throw he take from the line which set up a nice attacking play.' He placed a piece of chewing gum in his mouth.

Anil replaced the binoculars in its black case. 'I was in de Oval in 1996 when Latas score a hat-trick against Norway. De coach will be fired for not using Latas enough. Latas only play for about twenty minutes.'

Gabriel was angry. 'Maybe de damn coach did think Latas is 37 years old and would geh tired. Latapy is a smoker but he have the stamina to run more than all dem younger players and he is one of we best midfielders. He is ah gifted playmaker. Yuh didn't see how he was passing with accuracy.'

'Remember we have to blame Sancho for de own goal,' said Anil. 'Buh leh we look at de bright side. We is de smallest country ever to make it to a World Cup finals.'

Ming had a sullen look. 'So what is de big deal? Small or big, we should ah play better.' He thought about his recent job offer. An underworld figure, known as Chadee, offered him ten thousand dollars to be the driver in a kidnapping. He needed the money to assist with a mortgage but eventually declined the offer.

The West Indian fans in Germany still decided to have a street party. The victory for Trinidad and Tobago fans was not winning a game but being in the World Cup finals.

On the flight to Trinidad, Ming thought about the lavish reception that the team would receive. He wished he was a footballer on the national team. Aboud and Kervin were among the thousands of fans at the National Stadium who witnessed the members of the Soca Warriors receive their reward for an amazing performance. Each of the twenty-four members of the team received one million dollars which comprised $250,000 in cash and $750,000 in units from the Unit Trust Corporation. The prime minister announced the rewards given to other football players and coaches.

Aboud stopped cheering and clapping. He said, 'Wait na. One million dollars! Dat is ah lot of money. But dey geh all dat jus' for going to de World Cup! Giving dem de Chaconia Gold Medal is no big ting. Every dog on de street have medal or certificate. Buh dat is plenty, plenty money.'

Kervin shook his head. 'Dat is joke money think 'bout all de millions de government spend with the Soca Caravan in Germany. Why de government give $250,000 to de odder sixteen players who did not go to Germany? Money flowing freely now but hard times will soon come.'

A nearby fan joined the conversation. 'How de prime minister could reward de coach and de three assistant coaches and not give rewards to Justin? Is Justin who take money from he pocket to help de Soca Warriors. Is Justin who did speak to some of dem overseas players and convince dem to play for de country.'

'Leaving out Justin is pure politics,' said Aboud. 'Jus' because he is ah United People's Party supporter he geh sidelined. De public know de truth.'

Kervin nodded. 'We boys play good buh dey eh score a goal. Dey eh beat ah team. All dat reward was for jus' qualifying. Money really flowing in dis country. All we money from natural gas and oil going down de drain. Meh fadda use to tell meh 'bout de wastage of money during the oil boom years of de 1970s. De government use to spend as if dere is no tomorrow. History does really repeat itself.' After the function the elated crowd headed for their homes and apartments. They were satisfied and contented. The amount of money did not bother them. Heroes need to be rewarded.

Ming was a regular viewer of Watchers of Criminals, a daily programme which aired on Loser TV. The programme's host, Baldeo, highlighted crimes in Trinidad and Tobago. Victims often sent footage of criminal acts from surveillance cameras or their cellphones. Someone submitted a video of a car being stolen. The footage was aired on Watchers of Criminals.

That evening at 6pm, Baldeo, the host of the programme, appeared on television and said, 'Yes director I am here. Ah survive de assassination attempt. Director run de clip of de bandit.' After the short, fuzzy clip, the host continued.

'Ladies and gentlemen, it is alleged that this individual has been stealing cars for the past year. We will hunt him down! Look at de low-life, de good-for-nothing. Now he on national tv.' He slapped the table. 'And leh meh warn odder criminals dat I go mash dem up.' He snapped his fingers. 'Director doh hut yuh head, run de next clip when we visit de funeral home. Let's take a break. When we return is wetting after wetting.' Near his table was a mop and bucket which sought to reinforce his plans to wet criminals.

Most television viewers were eager to see the host visit a funeral home with relatives of a deceased woman. It was a morbid scene. The host wore gloves and a face mask. He pointed at stitched areas on the dead woman's arms and legs. He raised her arms so the viewers could see the wounds. 'Dese is chop wounds received from her attacker.' He paused and murmured, 'Hmm de body cold like dog nose.'

For five days each week, television viewers were treated to a comic performance deserving of an Academy Award. Some persons residing abroad listened to or viewed the programme on the internet. They loved the drama.

'Director run de odder clip from Pleasantville.' This was footage of a crime scene involving a shootout between the police and four bandits. Two of the bandits were killed. The host wore a bullet-proof vest and was surrounded by four tall security guards who were heavily armed.

'Hello, yuh live on Watchers of Criminals,' said Baldeo. His oversized spectacles rested on the tip of his nose.

'Yes, Mr. Baldeo dese Ming who is known as Shortman. Ah was in de shootout in Pleasantville.'

The host pointed at the table and camera then frantically waved his hands in the air. 'Yes, Mr. Shortman. Tell we de real story and what not. Wha' really happen?' He adjusted his spectacles and checked his cellphone. 'Allyuh tink I is ah chupidee?'

'De police shoot at we furst. All we do is defend we self. De police say we steal a car.'

'Ah believe yuh. Too much police corrupt and crooked and what not.' He slapped the table and pointed to the television camera. 'One is one. Director let's take a break.'

After three advertisements, the show resumed and the suspense continued. 'Mr. Baldeo meh two friends was innocent and now dey dead. We want justice and we want closure.'

'Can you identify de police, their car and what not?' He checked his cellphone and read two text messages. 'Can you identify de police?'

Ming paused. He was nervous. 'Nooooo…Mr. Baaaaldeo.'

'For de corrupt police out dere- doh chain up de people.' Baldeo began to sing and dance. 'Director, take dat in you waist. Sergeant Arnim and his team thanks for all yuh assistance. Remember to donate yuh ten dollars to this programme by texting FOOLISH.'

Ming regretted his criminal lifestyle. He was afraid of being caught and going to jail. Raphael had told him some horror stories of the Maximum Security Prison. He wished he had never stolen the car and robbed the beer trucks. He decided to maintain a low profile and avoid associating with his friends.

Marcus had built a small shack in the backyard of the family's home. He had applied to the Housing Developing Corporation for a new home. His application had been filed five years ago. He telephoned the clerk at the Corporation. She asked him if he was a supporter of the government and had a political party card. This was one of the major requirements if he wanted to be eligible for public housing. Marcus had membership cards for all the political parties in Trinidad and Tobago. He did not answer and hung up the phone.

He became disillusioned with politics when he learnt that the Jihadeen sect had made verbal agreements with political parties. Some politicians, deperate to win elections, made secret promises of land and jobs to this small sect. He was surprised to see that some members of this radical sect were given government contracts, involved in quarrying and were football coaches. Some businesses hired the Muslimeen as debt collectors. This was Trinidad and Tobago- a country with a short memory and where lawlessness prevailed. It was an era when the talented and innocent would be ignored. It was a time when the law-abiding citizens were thankful to be alive and government supporters were contented with their poverty-stricken state. Some said that Trinidad was not a real place. This was a country where the educated and skilled could be unemployed.

Yesterday he attended the funeral of Quasee who suffered a major stroke and had cirrhosis of the liver. Anil was an alcoholic and died after being in a diabetic coma. Marcus blamed the many beers they drank in 2006 for the demise of Anil and Quasee. Marcus and his friends disregarded their health and were concerned with winning the competition organized by Carib Brewery.

Marcus became disillusioned with society. His young wife had died of cancer because he could not afford chemotheraphy. She spent her final days in Vitas, a hospice in Port-of-Spain. His beard was long and his hair was uncombed. He had become a recluse and spent many hours composing songs or playing his cuatro. When he was a teenager, he played the cuatro for a parang group- Flores de San Jose. There was a knock at the door. He ignored it and heard a louder knock. He reluctantly opened it.

Aboud stood there with a look of shock and concern. 'Yuh hear dat Gabriel in hospital dying?' He placed one hand on his chest and struggled to breathe.

Marcus stared at him in disbelief. 'WHAT? YUH LIE! We Gabby? Nah, ah cyah believe it. Come inside and tell meh de whole story.' His attiude had changed.

Aboud entered and noticed the untidy room. There were old newspapers scattered on the floor. He sat on a sofa. 'Is true talk! De doctors and nurses say he have AIDS.' He saw a posey under the bed.

'WHAT? AIDS! When? How?'

'Must be when he was sharing dem dirty, infected needles from de pipers who does sleep in de Tunapuna market in de night.' Aboud looked at the ceiling. He did not want the tears to roll down his cheeks. He felt men did not cry.

Marcus briefly covered his face with both hands. 'Yuh sure is AIDS? Who tell yuh dat?'

Aboud steupsed and touched one of his ears. 'Ah hear it with meh own two ears.

Ah just meet Kervin now, now now… and he tell me dat he did faint three days ago and dey carry he to the Port-of-Spain General Hospital. Today ah went to see him and he had red rashes all over his body…a mere skin and bones. De nurses say he have serious diarrhoea. He very weak and coughing a lot. Gabby say that inside he mouth feel like some kind of moss growing.'

Marcus bit his lower lip. 'Dat could be AIDS symptoms. Doh be surprised if Jessika or Renata give him it.' He was infuriated. 'Dey too damn loose with themselves. De doctors sure? Because sometime when ah eat spoil food ah does geh a lil fever and diarrhoea.'

Aboud took a rag from his pocket and quickly wiped the tears rolling down his cheeks. 'Lissen to what ah telling yuh- is AIDS. It cyah be nothing else. Yuh remember when Gabby did tell we he had dry cough, flu symptoms and night sweats?'

'Yeah.' He touched his cheeks and realised it was tears. Both men were crying but were reluctant to comfort each other.

Aboud continued wiping his tears. 'Well he must have had it since den.'

Marcus had many fond memories of his childhood friend. 'Wha' ward he in?'

'Twelve.' Aboud folded his arms. 'Well, crapaud smoke he pipe. I eh goin' to no funeral. Ah fraid ah get bad name and people think I have AIDS too.

Let all he odder friends bury he.' He wanted to lighten the atmosphere. 'Thank God I ugly and pretty woman doh like me.'

'Yuh had no choice…yuh ugly…since yuh mudder and fadder ugly.'

Aboud smiled. 'Even if ah was good-looking, yuh go never ketch me in dat type of slackness.' He looked at the long line of ants on the floor. There was cobweb on the ceiling.

Marcus looked at the local fruits on the kitchen counter. 'Come leh we go and visit Gabby. We should carry some fruits for him. He like fig. He does eat plenty fig as if in he larse life he was a parrot. Maybe we could also carry some of dis pineapple.'

'Yeah. All dat sound good.'

Marcus placed some fruits in a brown bag. 'Come leh we go.'

'Yuh want to bathe or comb yuh hair? Ah go wait.'

Marcus patted his hair. 'Nah. Ah will do dat after de hospital visit.'

The hospital was a ten minute walk from Marcus's home. Both men continued the conversation whilst they headed for the hospital. Aboud had a black plastic bag.

Aboud said, 'If only Caribbean teenagers and young adults were more responsible then there would not be this AIDS epidemic.'

'One of meh cousins who working in de Ministry of Education tell meh dey did organize abstinence clubs in de secondary schools. This is a good weapon to educate we young people.'

Aboud nodded. 'In the United States there are many of these clubs and they have indeed had an impact on reducing AIDS. In my younger days we never hear 'bout HIV and AIDS buh we use to hear 'bout other sexual diseases as herpes, syphilis and gonorrhea.'

'Yes is true,' said Marcus. 'I really wish that the governments in the Caribbean would provide cheaper anti-retroviral drugs for the infected persons.The Caribbean needs more counsellors and therapists to provide emotional support. Every primary and secondary school should have a counsellor to advise and assist students. I eh know why all dem politicians want to put computers in de schools. Leh dem puh counsellors first and den puh computers.'

They reached the hospital and entered the main hall. Aboud informed the receptionist that they would like to visit Gabriel. She wrote their names in a thick notebook and pointed to a dimly lit corridor. Along the corridor were benches with elderly people who were moaning and shouting.

Marcus pointed the walls. 'Look at all these advertisements warning people about AIDS. And there are also ads for condoms especially near Carnival time.'

'Yes, I have noticed. We bombarded with ads about safe sex. So why is there still a crisis?'

'My view is dat sometimes when there is too much information on a subject, people could become desensitized and ignore the warnings.'

'Is true,' replied Aboud. 'Because people know the slogan 'Say No to Drugs' but are still involved in drug use and drug trafficking.' He opened the black plastic bag and found a large bottle of hand sanitizer. He was relieved and quickly liberally dabbed it on his neck, cheeks and both arms.

Marcus looked at him and laughed. 'Boy you think dat is perfume?'

'Man, I jus' trying to play safe. I don't want to ketch any virus or germs.'

They passed a noisy ward and peeped inside. It was filled with cribs and crying babies. There were five nurses busy feeding babies and changing diapers. Marcus asked a nurse, 'Is this the maternity ward?'

She was serious. 'No these babies have been abandoned by their mothers.'

'What are the reasons?' asked Aboud.

'They had birth defects like down syndrome, some have deformed limbs and others are perfectly healthy.'

Both men had shocked expressions. 'Right ok,' said Marcus. They quickly left the ward.

Marcus whispered to Aboud. 'I coming back here to adopt one or two babies.' Aboud remained quiet. It was a shock to see the abandoned babies.

At the end of the corridor they saw a doctor with a stethoscope around his neck. Marcus asked for directions to Ward Twelve.

They found the ward and quietly entered the room. Gabriel was asleep on a bed.

He was partly covered with a white sheet. His face looked pale. He had tubes in his arms and nose.

The ward had four other beds with elderly patients. There were two small tables with assorted medicine bottles and religious pamphlets and flyers.

'He alive?' asked Marcus. 'Look how magga he get!' The smell of disinfectant was everywhere. He felt nauseous.

'Yeah ah tink so. Shhh, yuh go wake him. Poor fella must be real suffering.'

Gabriel slowly opened his eyes and sneezed. 'Ah recognize dem voices. Marcus and Abu dat is you all?' He began coughing.

'Yeah is we,' said Aboud. 'We come to visit yuh and brought a few fruits.' He placed both hands in his pockets to ensure he did not touch anything. He quickly squeezed the bottle of hand sanitizer and rubbed some on his cheeks and forehead. He squirted some on hair.

Gabriel forced a smile. 'Thanks buh ah cannot keep anything down....' He sneezed and pointed to a green chair in the corner of the room. 'Leave de fruits dere.'

Marcus and Aboud quickly took a step backwards. They were petrified. Aboud placed the brown bag on the chair.

'So how meh family and Nigel, Ming and Anil?' He blew his nose in a napkin.

'Well...aamm...they are very concerned and worried about yuh health... in fact some of dem came looking for you but couldn't find de ward.' Aboud paused. He did not want to tell him the truth. 'A few of yuh cousins want to visit but yuh know dey doh have proper clothes to come and visit.'

Marcus added, 'Dey all badly want to see you.'

Gabriel coughed and cleared his throat. 'Come a little closer ah not hearing and seeing good.' His voice was becoming weaker and inaudible.

Marcus and Aboud cautiously moved closer to the bed. Marcus covered his nose with a handkerchief. 'So how de nurse and doctors treating yuh?' asked Marcus.

Gabriel had an expressionless face. 'Not too good. Everybody scorning meh just because ah have a bad flu. De nurse trying to sell me samples of some tablets.'

Aboud was shocked. 'What? She selling samples?'

'Dat is ah crime,' said Gabriel. He coughed and cleared his throat.

Aboud and Marcus quickly stepped away from the bed. Aboud opened the plastic bag and found a yellow towel. He placed it on his nose.

'Is ah bad flu dey say yuh have?' asked Marcus.

'Yeah.' Gabriel shivered. 'I think it getting worse.'

'Okay, we will come soon to visit again,' muttered Aboud. He quickly walked towards the door and noticed the dustbin that was overflowing with empty vials and injection needles.

'Right fellas. When ah geh better we go play some small goal football like de good ole days.' Gabriel wanted to wave his hand but he was too weak. His friends did not hear him.

Marcus nodded. 'Right, bye. And, remember doh buy no medicine samples and eat de fruits.' He was depressed. He thought about his dying friend and the abandoned babies.

Gabriel stared at the ceiling. He saw brown spots and assumed it was due to a leaking roof. He felt abandoned and lonely. He wondered if he would be going to heaven or hell.

His friends never returned. Before he contracted HIV, Gabriel was disliked by many persons because of his arrogance and boastful attitude. Two weeks after their visit to the hospital, Gabriel died. He had one child, Ritchie who was in Standard Five in Enterprise Government Primary School.

Marcus insisted that Gabriel wanted a burial. He told Gabriel's parents, 'How yuh could have a cremation for de boy? Gabby did always like flowers and he would have wanted a burial.'

Gabriel's parents refused to listen to the advice of Marcus. They did not attend their son's funeral service. They refused to let Ritchie be present at the service. Four persons attended his funeral and two persons were present at the cremation service. There was a stigma associated with AIDS.

The death of Gabriel transformed Marcus. He shaved his beard and hair. He began drinking bush tea and would spend more time by the river staring at the trees and fishes. He became withdrawn, more God-fearing and began to appreciate life. Furthermore, he wanted to repair relations with his father. He heard that his father would be walking along the shoulders of the highways searching for empty beer bottles. He would collect and exchange them for money at a recycling company in Champ Fleurs.

On Christmas Day there was a sudden change. Marcus wanted to renew friendships and invited eight close relatives and friends to a parang party at his home on Boxing Day. He was not sure who to invite. The lonely soul placed a parang cassette in the tape recorder and pressed the play button. There were local classics on the cassette including 'Anita', 'Soca Santa', 'Santa Looking for a Wife', 'Homemade Wine', 'Piece ah Pork' and 'Mamacita.' Whilst the music played, Marcus began to tidy the messy room.

Marcus knew he would never be the perfect father but he wanted to be a good provider for his son and adopted daughter- Kyle and Carla. His daughter had been abandoned at the hospital. In 2002, he carried both children to a rally in San Fernando. The event was organized by the Ethiopian Peace Foundation to discuss the visit of the grandson of Haile Selassie 1 of Ethiopia- Zerah Yacobe Asfa Wossen. Selassie's grandson was to visit the island during 27-30 November 2002. It was a memorable event for Marcus and made him more conscious of his African heritage.

In 2004, Marcus attended a function at Motown Pan Theatre on Navet Road in San Fernando. At the function, awards were given to Rastafarians who made outstanding contributions to society. Marcus became sentimental when he walked on the stage to receive a sporting award on behalf of Emmanuel. After the ceremony he was able to meet some of the honourees including Roy Cape, Brother Resistance and Black Stalin.

Marcus was interested in his children's education and would attend the annual Parent's Day at Presentation College in Chaguanas and Holy Faith Convent in Couva. Both were prestigious Roman Catholic schools in central Trinidad. Every evening, except weekends, he would assist his children with their homework and paid for their extra lessons on Saturdays. He hoped both children would win government scholarships.

Marcus entered the bedroom of his two children. He told Kyle, 'Study hard while you are at school. Doh waste yuh time on de internet yuh see all dat Hi5, e-mail, Facebook, Twitter, My Space or having a blog…all dat is North American garbage. Yuh wasting yuh time.'

Kyle smiled. 'Yes okay, choo dat.'

Marcus continued. 'Yuh have to resist mimicking yuh friends. It go start up small when you see dem with ah earring in dem ears and you geh one.' He sneezed. 'Den before yuh know it, you have earrings and borrowing yuh sister lipstick and high heel shoes. De society changing but you don't have to change.'

Kyle and Carla laughed.

'Be a leader not a follower.' He had similar advice for his daughter. 'Doh worry 'bout boyfren and fashion. You still young. Yuh beautiful but don't enter beauty contests. Study yuh books. Focus on yuh studies and get a good job. And when is time to choose boyfren doh choose a saga boy who will grow up to be a macho man. Choose someone who is educated, will treat yuh good and be a responsible parent. If yuh choose wrong yuh will ketch yuh nen nen.'

Carla nodded. 'Yes daddy, of course.' She never thought of her father's advice as old-fashioned, outdated or traditional. She had a bowl with governor plums.

'Now both of you lissen well.' Marcus sat on the edge of Carla's bed. 'You all might geh a chance to study abroad buh remember to return home to contribute to de development of dis country. If dem country bookie and others want to go abroad leh dem go.' He sneezed. 'It simply mean more jobs for de rest of we and less traffic on de roads. Some Caribbean people…like dem country bookie, rather leave here and go in a strange land and face coldness and do three jobs and leh foreigners spit in dey face. Be proud of dis land. We win Olympic medals and produce Nobel Prize winners. De fastest man in de world from Jamaica. De Caribbean is de best. Yuh understand?'

The atmosphere was tense. Both children nodded. They were serious.

Marcus sneezed. 'Ah feel dis Baygon cause meh to sneeze. All dese repellants and burning dis cockset, bugmat, vapemat and citronella candles not good for all yuh to inhale. It go cause cancer later on. When ah was younger meh fadda and grandfadda always using Flit and it did work good. It might be better to have nothing burning and spray nothing and take a chance with de mosquitoes.'

Carla quickly replied, 'Buh daddy, mosquitoes does cause dengue and people does sometimes die with it.' She looked at her brother's face. He wanted to laugh but decided to remain quiet.

Mosquitoes were prevalent in the neighbourhood and there had been cases of persons who died of dengue. The Ministry of Health occasionally despatched vans and health inspectors into communities to spray the bushes and drains. The spraying was undertaken at 5am or 6pm. The spraying contributed to a decline in dengue cases but there was soon an increase of patients with respiratory problems. The Ministry also issued media releases stating that there were no reported dengue deaths in the hospital. This was a public relations gimmick undertaken by the Ministry so as not to alarm the public. The Minister of Health informed doctors and hospital personnel to hide files of patients diagnosed with dengue. The Minister did not want the media to know the truth. A few doctors publicly revealed that their patients were diagnosed with dengue and this caused panic.

'Now two of you go and play outside. Doh leh de mosquitoes bite yuh. Some of meh friends coming next hour to rehearse for de parang group.' Marcus wondered if they would remember his advice on careers and studying. He moved the old Sunday newspapers. He glanced at the headlines on the back page. There was a report on the finals of the World Cup. He was not interested and watched the clock on the bureau. It was a birthday gift from his sister. He was anxious to meet his friends. Each had a nickname. Delisa was known as Parasite because she always boasted that the government provided handouts and housing for her family. She had small eyes and a nose that resembled a snout. She was asthmatic and epileptic but refused to take medication. Wilfred was known as Mango because his head was shaped like a Julie mango. Then there was Bertrand who was known as Fig. He was given this nickname because he had seven plantain trees in his yard and his mother had an old, fruit stall. These friends also had two nicknames for Marcus- Black Pudding and coal pot dougla. They gave him those names because of his dark skin and also because his favourite food was Charlie's black pudding and fresh hops bread.

Bertrand arrived an hour earlier and was discussing cricket with Marcus. Delisa soon arrived with her six year old son, Aaron. She was eager to join the conversation. She wore a short pants and a tight green jersey. She had recently been in prison for physically abusing her young adopted daughter, Alanay. The doctors claimed it was postpartum depression but this did not convince the jury.

'Aye Parasite what is de latest movie you see dis week?' asked Marcus. He knew that she regularly visited the cinema and patronised the DVD stores.

'Oh, Princess Diaries 2. It was oh so wonderful. Wha' you tink Bert?'

Bertrand imitated her voice. 'Ooohhh, yes Parasite. It was wicked! Loved part two. How I wish I was living in Genova!'

Delisa did not seem to mind the mocking. She felt this was a sign that he liked her. 'Did you see Legally Blonde 2? It was better dan Troy and Titantic.'

Bertrand smiled. He was accustomed to her eccentric behaviour and odd remarks.

Wilfred quietly entered the room. 'YES. Ooohhh, de little dawg so cute,' he said in a mocking voice. 'Dat show was better than part one and I cried through it. Parasite yuh know I'm so sentimental.'

He was greeted by Bertrand. Both men recently spoke about attending a cricket match. Bertrand hugged him. 'De doods man himself! Long time no see.' He extended his arm and had a closed fist. 'Give meh a bounce na.'

Wilfred was the owner of a thriving funeral home. The high number of murders partly contributed to the financial success of his business. He hired Bertrand to steal coffins from cemeteries. It was a morbid job but they nedded the money. Bertrand would check the death notices and attend funerals. He would be at the gravesite mourning with the friends and relatives of the deceased. Later in the night he and an assistant would return with shovels and unearth the coffin. He would throw the dead body in the grave and shovel soil into the hole. Upon receiving the coffin, Wilfred would wash it and re-sell it to new customers. It was at the Tunapuna Public Cemetery that Marcus, who visited gravesites to steal flowers, met Bertrand. Marcus was no longer involved in stealing flowers and both men remained good friends.

Marcus stood at the door of the kitchen 'Aye Mango what's up? Meh hands dirty ah putting some cloves in de ham. Pull up a seat.'

'Okay boss,' said Wilfred. He sat down and crossed his legs. He lit a cigarette. 'Ah waiting long time to eat some ham and hops.'

Bertrand licked his lips. 'Yeah, ham is my favourite.' He gently tugged at the stubble on his chin. He did not shave for four days.

Wilfred looked surprised. 'Buh yuh surname is Mohammed. Dat means you is a Muslim. So what you eating pork for?'

'Dis not pork is ham!' replied Bertrand. He swallowed and continued. 'Somebody surname have nothing to do with race and religion.'

Marcus smiled. 'Yuh right about surnames. But what you think ham is? Where yuh tink it come from?'

Bertrand bit his lower lip. 'Hmmm…I always thought ham meat come from hamsters.' He smiled. 'Buh wait na, ah realize dat meh fadda was ah Muslim and mudder was a Hindu so I is half Hindu and half Muslim, so is ok if ah eat ah lil piece.'

Wilfred jokingly said, 'Buh if you is a real Hindu, you supposed to be a vegetarian, so how you eating ham, beef pastelle and drinking alcohol?'

Bertrand did not reply and had a broad smile. He was an atheist but not reveal this because he feared being mocked by his friends. He was bored with the discussion.

Marcus left and returned with a bowl of vegetables and a small knife. 'Will, yuh always have to make tings difficult.' He paused for a moment. 'Well de turkey was a vegetarian when it was alive, so is okay to eat it. Remember also hamsters are vegetarian so is also alright if he eat ham.' He began to hum 'Neighbour, Neighbour' which was an old parang sung by Rikki Jai.

Wilfred laughed. 'All yuh full of stupid talk. Anyway this is why Chrissmass is special and different from other religious celebrations. I know some Hindus, Muslims, Jews and other non-Christians who celebrate it.'

Marcus joked. 'Yeah even cunumunu people like you could celebrate it.' He began to cut onions and celery.

Bertrand and Wilfred laughed. Bertrand said, 'Is true. Chrissmass morning, Hindu and Muslim children getting gifts under their Chrissmass trees and their parents also light up de house with lights and ting. Lissen this morning on ESPN had a documentary on some of the world's greatest goals. In one game of de English clubs, the goalkeeper from de opposing side came forward when a corner was being taken. And dat ball fall directly on the keeper's foot who rock it into the back of the net.'

Wilfred responded, 'Furse time ah hear a goalkeeper scoring during a game.'

'Yeah, usually the keeper has a chance to score when is penalty kicks,' said Marcus. He was cutting onions and peeling pieces of garlic. 'Twice ah saw a striker score by simply chipping de ball over the head of the keeper. De keeper had come out of his area for jus' a moment and he paid dearly. In one game a striker slip de ball between de keeper's legs.'

'Yes de ESPN programme had two goals like dat,' said Bertrand. 'Dere was another beauty of a goal dat was scored outside the penalty box. De striker fire a shot and caught the defence and keeper totally off-guard. Dat ball move like a rocket and it hit de inside cross bar.' He excitedly waved his arms. 'De programme should have shown a Black boy from London- Paul Ince. He played for England and one time he did an overhead kick to score a spectacular goal. Dey had another Black who was born in Suriname and played for Holland. He name is Seedorf and is a good midfielder.'

Marcus stopped peeling garlic. 'Yes, ah know Ince. He was from de East-End in London. He was a dangerous marksman for goalkeepers. One time Ince receive a volley and score with a sidekick. Ah doh know Seedorf.'

'Seedorf play in Euro 96,' replied Wilfred. 'All yuh does follow de Primera División of La Liga?'

Bertrand's face was animated. 'But of course! Since ah small ah supporting teams like Real Sociedad, Athletic Bilbao, Valencia, Deportivo, Sevilla and Real Betis. Dese days ah like the usual favourites, Barcelona and Real Madrid.'

Marcus asked Bertrand, 'Remember the Copa del Rey final with Barcelona versus Real Madrid?'

'Who could forget dat! Yes it was great. Did you see the first Beach Soccer World Championship in 1995 at Copacabana Beach in Rio de Janeiro when Brazil won?'

'Of course. Ah geh interested when ah see football stars as Eric Cantona, Zico and Julio Salinas participating in these games. Ah remember as a little boy playing football on Maracas Beach but dat was for fun not competition.'

Wilfred added, 'Yeah I play in Maracas and Mayaro. Dis beach football really develop quickly all over de world. Look it had de Pro Beach Soccer Tour in 1996 and two years later there was de European Pro Beach Soccer League. It seems dat de formation of de Beach Soccer Worldwide is an attempt to make it a strong body like FIFA. Did you….'

The conversation was interrupted when the doorbell rang. Bertrand glanced at Wilfred and opened the door. 'Yeah how could I help you?'

'Blessings to de brethren and sistren. I am Mala and this is Sister Sunita,' said Mala. 'I saw de tent outside and knew it was a party. Could we attend dis party? What is de cost?'

'Dis is a private party.' He did not know the women and began to close the door.

Mala spotted Delisa in the hallway. 'Wait. Delisa dat is you gyul? How tings?'

Delisa was startled to hear her name being called by a stranger. She cautiously approached the front door. 'Yes. Who are you? Do I know you?'

'Is me Mala. Two years ago you called me by mistake on the telephone and den we later meet in the mall. Well dis is Sunita and her son Sunil. Dey jus' return from de United States.'

Mala was infamous for getting into private parties. She saw this party as an opportunity to obtain a free dinner.

'Amm. A wrong number. Okay.' Delisa was alarmed.

'Do you have gift paper and a scissors?' asked Mala.

Delisa looked at Bertrand and then at Mala. 'No. Dis is not my house. Why do you need it?'

Mala had a headache. She felt it was due to the heat. 'Oh, well I have two Christmas gifts and a birthday gift for you and de family, and wanted to wrap it and bring it later for you.' It was an odd comment since it was the month of May and Mala did not know the date of Delisa's birthday.

Sunita decided to tell a convincing story which would allow them to enter and be accepted in the party. 'We from de Hare Krishna group in Enterprise. I collecting money fuh de temple and if yuh have foodstuffs to donate. We hope tuh buy groceries and create charity bags to distribute to the poor and underprivileged. All yuh cud assist?'

Delisa smiled and told Bertrand to allow them to enter. She told the group, 'Dat sound so nice. Here's some jewelry and Revlon and Sacha cosmetics. I don't understand why only at Chrissmass time people does want to assist de poor.'

'Our group is more concerned with food.' Sunita was agitated.

Bertrand intervened. 'Parasite put away your make-up. Wha' hungry and poor people will do with yuh stuff?' He had a serious expression. 'Throughout de whole year, dey ignore de orphans, de vagrants, disabled, squatters like we and de sick in de hospitals.'

Delisa took a small mirror from her purse and stared into it. She added lip gloss and used a small sponge to pat her forehead. 'Hold up selector, yuh trying tuh diss de Hare Krishna program. Is not only dem doing dis. For ah long time de whole society like dis.'

'Parasite stay out of this. Come in de porch women and leh we talk some more.' Bertrand had a stern expression. He looked at the crotons in the rusty pots. They needed manure and watering.

Sunita and Mala obeyed. They entered the crowded porch and sat near the small centre-table. Sunita's son, Sunil, also entered and sat near his mother. On the centre-table was a platter with fruit cake, candy canes, nuts, pastelles and biscuits.

Bertrand said, 'Everybody is a bunch of hypocrites. We feel dat giving de poor a bag of groceries and some presents for one day will give dem de Christmas spirit.'

Sunita asked, 'So when de groceries done dey Christmas spirit gone too?' She gobbled up a large piece of fruit cake. There were flies on the windows. The window frames were rusty.

'Of course,' Bertrand replied. He had a slice of spongecake in one hand. 'Some people like to play dey so charitable and loving during Christmas but for de next eleven months dey pass dese homeless people straight. Most of de society only concerned about demselves. Marcus bring some groceries for de Hare Krishna woman.' He sipped the gingerbeer. Flies buzzed near the table and two landed on the biscuits.

He steupsed. 'Yeah leh meh wash meh hands and ah will pack some tings.'

Delisa looked at Sunita. 'What is de true meaning of Christmas?'

Sunita was hesitant to answer. She looked at Mala who seemed agitated and had a perplexed expression.

Mala regained her composure and responded, 'We aint Christians buh we know Christmas is not only about shopping, gifts and Christmas trees and Santa Claus. We have to remember the birth of Jesus Christ, our Lord and Saviour. Anyone who feels dat Christmas is only about parang, poinsetias and pastelles, they are like de Grinch who stole Christmas.' She paused to eat a pastelle. 'Anyone who feels dat the season is only about cleaning, painting house, changing curtains and baking Christmas cake, they are like Scrooge. Once you remove 'Christ' from Christmas you remain with one word – MAS, and we know mas is dat annual festival of stabbings and promiscuity - Carnival.'

Bertrand looked surprised. 'Dat is an excellent statement…except for de part 'bout Carnival. Buh yuh is not even a Christian and yuh know about the true meaning of Christmas. I think it have a song called 'Put Jesus in Yuh Christmas' buh ah cyah remember who sing it, might have been Crazy.'

Delisa shoved the mirror, sponge and lip gloss in the purse. 'It easy to talk about de true meaning of Christmas but nobody will take you seriously.'

Wilfred was seated nearby in the porch. He jokingly shouted, 'Mrs. Delisa Rampersad…you need some serious help. Have you seen a psychiatrist recently?'

She nodded. 'Actually I am getting help from a professional with integrity and morals who by de way was a former criminal.'

Wilfred and Bertrand laughed. Mala and Sunita smiled.

'Yuh real brass face,' said Bertrand. He asked, 'How come anytime we see yuh, is ah different wig yuh wearing? De government giving yuh a wig allowance or a weave allowance? How yuh could afford dat?'

Delisa patted her head. 'Doh tell anyone buh ah in ah ghost gang in de correct Unemployment Relief Programme. Ah using names like JLo, Kardashian and Beyonce to collect three salaries every fortnight. Recently ah found some bogus hundred dollar bills.'

'Did you carry them to the police?' asked Wilfred.

She patted her chest. 'Who me? Ah might be poor but ah not stupid. If I give it to the police they will spend it.'

Bertrand said, 'Dat corrupt Unemployment Relief Programme is ah waste of time. Larse month dey had a big fight outside meh home. Dey was fighting over who could buss more stone. All dem workers stealing people fruits and does stop every fifteen minutes to ask for water or smoke cigarettes.'

Marcus entered the porch. In his right hand he had a bag filled with canned foods and powdered milk. 'Season's greetings.'

Mala and Sunita clasped their hands and bowed their heads. 'Season's Greetings brother.'

Marcus looked confused. 'Do I know both of you?' He gave the bag to Mala.

'We is Delisa friends,' said Mala who placed the bag on the ground.

Delisa rolled her eyes. Her cellphone rang and she answered it. 'Hello, yes. That's wonderful news, congratulations. Really? That's amazing. I'll call you back this evening. Bye.' After ending the conversation she placed the phone in her purse. 'Allyuh meh cousin jus' had triplets in de hospital. Two boys and a girl chile.'

Mala was the first to respond. 'Congratulations!' She took a handful of nuts and shoved them in her mouth. She began to furiously chew and then bit into a slice of cake.

Bertrand and Wilfred were surprised at the greediness of Mala. Bertrand shook his head in disbelief.

'Congrats!' said Marcus. 'Did she give dem any names as yet?' He stroked his goatee that was recently dyed.

'Yeah, but ah not sure. She say dese three babies will get nuff respect. Dey born with hair on dey head.'

Sunil laughed. His mother gave him a stern look. He took a candy cane from the platter and began to chew it. Mala and Sunita listened attentively. Both women hoped they would receive an invitation to stay for dinner.

Delisa looked at her wristwatch. 'Lissen honeys, I have an appointment wit de hairdresser, Pablo. He is going to put a relaxer in my hair. After dat ah have to go also for my facial, to geh my eyebrows plucked and legs shaved and a pedicure and manicure.' She began to bite her nails. 'Ah leaving meh son here awhile and will pick him up tonight. Tomorrow is work for me.' She worked as a waitress at Villa Capri in Marabella.

'Just go nah. Yuh doh have to leh de whole world know yuh business,' said Wilfred, 'yeah we go keep an eye on the lil man.' He patted Aaron on his head.

'Yuh stop visiting Eddie in jail? How he going?' asked Bertrand.

Edward was her boyfriend and father of Aaron. He had been sentenced to serve six years for armed robbery. He hoped that after being released he would obtain a United States visa. He wanted to begin a new life in North America.

Delisa quickly replied, 'Yeah ah cut down on visits since ah give Eddie a BlackBerry cellphone. We in touch regularly. With his phone, he in touch with all his gangsta friends from de ghetto and homeys from de hood. Eddie also using his phone to convince two witnesses not to testify against him.'

Wilfred had a shocked expression. 'Ed have a cellphone? How de hell dat could 'appen?'

'I sneaked it in de prison 'bout four months ago. Ah prison officer have a charger for him to charge it up anytime de battery low. Having a phone eh no big ting. Other prisoners have upgraded their cellphones and some even have internet access. Dat is how some of dem prisoners could still continue calling de shots on rivals and de police. One prisoner even own de latest iPad and Kindle. Anyhow, bye boys and girlfriends.' She quickly exited the house and closed the gate. She headed for the nearby beauty saloon.

Bertrand was temporarily dumbstruck. 'Imagine prisoners have a better cellphone and technology dan me.' He turned and asked Aaron, 'Small man yuh want anything to drink?'

Aaron meekly replied, 'Yeah sorrel and cake.' He was sucking his thumb. On his lap was a plastic bag with marbles.

Bertrand quickly replied. 'I will get some unsweetened sorrel for you but no cake because of de diabetes. And then we go find a cartoon for you to watch on tv.' He swatted a fly that landed on a small table.

'How ah lil boy like dat geh diabetes?' asked Mala.

'Is all dem free box lunches de chirren getting in de school-feeding programme. It does have plenty rice, macaroni and other carbohydrates. Some of de chirren does eat two and three boxes for lunch. And some teachers does beat de chirren if dey cannot eat all de food.' Bertrand departed for the kitchen. He did not like discussing sickness and disease.

Mala looked at the kitchen. She smiled and asked Wilfred, 'So is dat duck and dhalpuri I'm smelling?'

Wilfred replied, 'Yeah buh not here. It coming from de neighbour house.'

'Could I geh some food in a container to carry home for meh husband and children?' asked Mala. 'Doh take it ah how.' She coldly stared at him.

Her brazenness shocked Wilfred. Marcus had grown accustomed to her flippant behaviour. He smirked and lit a cigarette. He threw the smouldering matchstick in the grass.

Marcus quickly responded, 'We didn't cook enough to share.' He cleared his throat and stared at the gate. 'Woman yuh is ah real boderation.'

Sunita felt uncomfortable and wanted to leave and visit other homes. 'Do you have any money yuh want to donate for de poor?' She tried to signal to Mala that now was an appropriate time to depart.

Wilfred yawned. 'Breds yuh come too late, ah already spend meh money on tickets for de concerts featuring Richard Marx, K.C. and de Sunshine Band, Beyonce, Chippendales, and some Bollywood group in ah Unforgettable Tour.' He looked at Marcus.

Marcus shook his head. 'Ah scrunting. Every month I does have to pay water and electricity bills. And it is plenty since I have eleven air conditions in de house, seven showers, two swimming pools and have to wash regularly. Ah brokes, ah cyah even afford to go to Chutney concerts.'

Marcus glanced suspiciously at both women. He did not like the presence of the two uninvited persons. He felt Mala resembled one of Gabriel's former girlfriends.

Bertrand returned with glasses of sorrel and gingerbeer and three packs of Dixee crackers. He placed the tray on the small centre-table. 'Allyuh help yuhself to de drinks and biscuits.'

Both Sunita and Sunil took sorrel. Aaron decided on gingerbeer. His bandaged hand was paining but he refused to complain.

'Buh allyuh cannot spare some money to assist de less fortunate?' asked Mala.

She took a glass of gingerbeer. 'Remember Jesus Christ was born into dis world to bring happiness; to bring peace and goodwill to everyone....'

Bertrand interrupted. 'Lissen, I have to pay cable, telephone and internet bills. And of course- my cellphone bill. Wilfred give some money nah. You always have excuses!'

'When? Yuh farse and out of place…when? Tell meh!' Wilfred was angry.

'Will doh geh aggressive. Earlier dis year, some people was collecting money to help build a room in an orphanage and yuh tell dem you would ah help later buh yuh never did.' Bertrand sipped his drink. He focused on the song from the neighbour's radio. It was sung by Luther Vandross. He could not recall the name of the song.

Wilfred remained pensive and then quietly responded. 'Oh yeah, ah had to save money to go to concerts by Sugar Ray, Michael Bolton, Lionel Richie, Air Supply, Scorpions, Nikki Minaj, Evanescence and R.E.O. Speedwagon.'

'Well thank you for your time. We have to visit other homes,' said Sunita.

Sunita, Mala and Sunil left the compound and went to the neighbour's home.

Wilfred returned to the living room. 'Two of dem is real tricksters. Dey both need a good cutarse. Lissen fellas ah feel ah want to go home and relax, ah had a hard day at work today.' He peeped through the curtains. 'De rain set up.'

Bertrand nodded. 'Ah already size dem up long time. Marcus ah feel ah want to go home and relax a bit. Leh we arrange for another day.'

Marcus shook his head 'Yeah all yuh fix up na. Next time ah having ah party or lime ah will hire a security guard. Ah have to fix up de house, meh two sisters and their children coming Thursday.' He planned to vacuum the carpet on Wednesday. 'Ah eh able when family visiting.'

On Christmas Eve, Cleopatra and Harriet arrived at the home of Marcus. They planned to spend three days at this residence.

Cleopatra hugged her brother and entered the living room. In the corner there was a potted plant decorated with blinking lights. At the base of the tree were four boxes wrapped in gift-paper and there was a small crèche with figurines. The living room had two couches and a broken rocking chair. Parang music filled the air. She began humming 'Ah Drinking Anything.'

The next morning, Cleopatra awoke early and decided to clean the living room and kitchen. She placed a stained apron around her waist and began to scrub the kitchen sink. After fifteen minutes, she switched on the radio and opened the freezer. She wondered if Marcus had purchased pastelles. She checked the refrigerator but did not see any pastelles and ponche de crème. She listened to the raindrops on the roof.

Malcolm and Raphael were expected to visit in the morning and spend the day. Harriet was still sleeping. The house was cool and this was due to the shower of rain.

Cleopatra asked Marcus, 'You remember when we was smaller how Mammy used to have me and Harriet cleaning house for Miss Bhagwoutie… and…odder ole people?' She was reluctant to ask the question.

Marcus stopped reading the newspaper. He folded it and placed it on his lap. 'Yeah. How ah could forgeh dat? It use to always cause confusion and bad blood in we house. Mammy even force daddy to change de ole woman will so we go geh de house and land.' He placed one hand under his chin and looked at the ceiling. 'All we cousins and neighbours use to laugh and bad talk we. Twice Bhagwoutie chirren send police for we. De police say we kidnap de ole woman. You and Harriet almost make a jail for tiefing de woman furniture.'

Cleopatra looked forlorn. 'Ah does still geh nightmares. It does still affect Harriet. It wasn't right. Mammy had we tiefing from de old people. We was like corbeau. We had to lie and tell people we winning tings in competitions. Buh dey never see competitions advertised or de results in de newspapers or television so dey know we was lying. People did laugh when bandits rob de store ten years ago and force Mammy to close down de shop.'

'Thief from thief make God laugh.' Marcus pointed to an old couch in the living room. 'Dat come from Mister Gopaul de retired principal who had all he chirren living abroad. Instead of bathing him, Mammy told us to hose him down.'

She wiped a tear that was trickling down her cheek. 'Today dat is considered elder abuse.'

He continued. 'I use to dread connecting dat water hose and carrying him outside to wash him down…like some animal. Even now whenever I use de powerwasher during Chrissmass ah does remember dem dread days.' He then pointed to a large vase. It had a crack at the top. 'Dat belong to Miss Chesney.' He sighed. 'Mammy give meh it. Dat air condition and dis fan come from ole lady Williams and de broken parrot cage from Uncle Neveal. One of dese days ah going to throw away all dem items. Dem tings like ah blight in de house.'

Cleopatra nodded. 'Yes, ah did have to cook for Gopaul and he did have a mentally retarded son. Everybody use to laugh at Gopaul because he use to pull his pants as high as possible till it was halfway up his chest. Ah never work for Chesney. Harriet did work for she. Harriet did have to bathe she and cut she toenails…things we never do for we own mudder. Mammy use to take Chesney pension money and cash it and keep most of de money. One time I had to shave she chin and cheeks…it had hair growing.' She stared at the radio and wondered if it belonged to a stranger.

Marcus looked at a painting on the wall. It was by an American artist and depicted a river scene. 'We have to forgeh 'bout de past buh make sure we chirren never end up so. Dere is something call generation curses where de sins of de parents fall on de chirren.'

'Dat easy for you to stay. Yuh was never involved like we. De past never really go away, it stays with you and yuh chirren.'

Marcus quickly replied, 'I was involved in de scam. Remember sometimes ah did have to steal flowers from de cemetery late at night. Is dere ah furse meet Bertrand. Remember when ah went in jail for six months for stealing white orchids for Mammy? Like yuh forgeh? Yuh know how much customers use to complain to me dat dey geh rip off with de floral arrangements. Mammy use to take all de dried and dying flowers and hide dem in the middle of de arrangements.'

Cleopatra wiped the tears that were rolling down her cheek. 'We better forgeh 'bout the past for now. People must wonder if we was one of the forty thieves with Ali Baba.' She continued cleaning.

Marcus opened the newspaper and began reading the section on international news. Three soldiers died in a shootout in Israel. He wondered if there would ever be permanent peace in the Middle East. There was a train accident in China. He turned to the sports section and checked the listings of European football leagues. In the German League, his two favourite teams-Wolfsburg and Stuttgart were near the bottom. He was glad to see Inter Milan, Napoli and Juventus in the first three rungs of the Italian League. In the Scottish League, Celtic was at the top and Hamilton was last. On page nine was a large advertisement paid for by the outpatients of the St. Ann's Psychiatric Hospital. The advertisement informed the public of the group's Annual General Meeting and election of persons to fill the posts of president, secretary and treasurer. The outpatients referred to themselves as alumni and offered scholarships to anyone who wanted treatment at the hospital.

On the nearby table were jugs with gingerbeer and sorrel. There was an old mop and broken radio which Marcus planned to dispose of in the nearby river.

Justin, Harriet's son, was nine years old and questioned the existence and ability of Santa Claus. He wanted to view Sesame Street but decided to watch a cartoon about elves residing in the North Pole. He asked, 'Uncle Marcus how could Santa Claus give gifts to all de children in de world in one night?'

Marcus momentarily stopped reading the newspaper. He smiled was not prepared for the question. 'Well, some people are hard workers. Take for instance, the Honourable Minister of Education in this country. Every morning for five days a week she wakes up early in de morning to cook thousands of breakfasses and lunches for school chirren all over de country. And de bright prime minister….well poor thing…he have to get up even earlier to go to de gym.' He stopped and sipped his coffee. 'Dem is real heroes, just like you and me.'

Harriet slowly descended the stairs. She was moaning and had one hand on her head. 'Merry Christmas everyone. Justin, finish help clean de house, yuh mudder too weak and tired to do anything. I'm going to sap meh head with some brandy and ah going back to sleep.'

Marcus stopped reading the newspaper. 'Gyul you go and rest. For de larse week yuh killing yuhself painting fence, scrubbing moss from de terrazzo and hanging icicle lights on the roof. Last year yuh break yuh foot when yuh fall off a ladder. Tomorrow yuh go complain yuh back or head hurting. For some people only if dey tired den dey does feel de place clean and geh de Christmas spirit.'

'Yes you go back to sleep and we will take care of everything,' said Cleopatra. She felt weak and this was due to her low blood pressure.

There was a knock at the door. Cleopatra placed the broom next to the dining table and peeped through the window. She opened the door and mumbled to the persons outside.

Justin was inquisitive. 'Who is dat Aunty Cleo?'

'Doh worry dat was some beggars.' Cleopatra walked towards the kitchen. She quickly wiped her hands on an apron.

Justin followed her. He was eager to see the cooked ham.

'De ham smelling like it finish.' She cautiously opened the oven. 'Yes it done.

Ah going to put de turkey in now.' She placed the ham near the sink and adjusted her apron. 'Later ah going to make some beef and chicken pastelles, fruit cake and ponche de crème.'

Cleopatra removed the liver, heart and gizzard from the turkey and placed them in an enamel bowl. The giblets would be used in the stuffing which was a favourite of Justin. Last Christmas Eve there was a big rat living in the stove and during the cooking of the stuffing it ran out into the dining room and frightened the guests.

She looked at Justin who was poking the ham. 'Justin leave dat alone. If you were a ham would you like someone poking you?'

'No aunty. I jus' checking to see if it cook.'

'Where yuh fadda? He home?'

'He gone away for awhile buh he still like mummy. Daddy does give mummy plenty compliments. Larse month on mummy birthday, daddy told her dat she big and fat like a barrel and smell like rotten salt fish.'

Cleopatra was embarrassed. 'Okay. So how school?'

Justin replied in a sad voice. 'Ah geh suspended for two weeks from school.'

'Wha' for? You is ah good chile.'

'Yes I know I good. Yuh didn't have to tell me dat. De principal suspend me because ah brought a pitbull dog to school.'

Cleopatra opened her eyes. 'WHAT? You mad boy!'

Marcus overheard the conversation and decided to intervene. He entered the kitchen. 'Justin yuh real skylarking in school. What de hell yuh doing with a pitbull?'

On the radio there was the parang 'Hush Yuh Mouth' by Kenny J. Marcus lowered the volume. Justin was afraid of further scolding.

'Well, ah bring it for two friends who never see one. Den it escape and bite a girl and de security guard had to shoot it.'

Cleopatra looked at the roof. 'Oh Lawd! Ah cyah believe dat! Just yesterday Dog play in de evening for Play-Whe.' She tugged at her hair. 'Damn ah would ah be rich.'

Marcus smiled. 'Justin so if de teacher did talk 'bout Noah's Ark, yuh would ah bring ah whole zoo to school? When ah was your age ah use to spend my time playing cricket or kicking ball not playing with pitbull. Yuh mudder mentioned dat yuh took part in the Algico/Rotary Games and de Milo Games. If yuh doh like book den focus on running in school. Do something constructive or else yuh will end up in the Youth Training Centre. Yuh want dat?'

'No, Uncle Marcus. I want to finish school.' The conversation was interrupted. There was a knock at the door.

'Shhh, ah hearing knocking on de door.' He peeped through a window. 'Oh damn is de police!' Marcus whispered to Cleopatra. 'It could be thieves and kidnappers who disguised as police. It's a normal ting in dis country. Let's all be quiet and dey will soon leave.'

'Uncle Marcus why we not opening de door for de police?'

Marcus quickly placed his index finger on his lips. He did not want to tell him the truth. 'Sshhh. De want to lock we up for baking ham and turkey. In this part of de country cooking dem ting is illegal.' He smiled.

Chapter 7
Temporary spectators and a tourist

On Saturday evening Wilfred, Justin, Ritchie, Kyle, Bertrand and Marcus participated in a windball cricket game on the Maracas beach. It was a fun game and referred to as a fete match. The highlight of the game occurred when Ritchie was batting. On two occasions the ball hit him on his head and sped to the boundary. The other players laughed when they realized he had made eight runs without a single stroke. Wilfred adjusted the dial of a new radio he had purchased. It was his first radio. It was metallic green with a short antenna and could easily fit into his pocket. He was eager to witness and listen to the upcoming International Cricket Council World Cup match between the West Indies and Australia at the Queen's Park Oval in the city.

Next day, Kyle, Wilfred, Marcus and Bertrand waited at the bus stop. They were heading for the cricket game. They arrived early despite encountering traffic and hurriedly headed for the stadium's entrance which had a long line of season ticket holders. There were four unkempt scalpers selling tickets outside the Oval.

'Ah cyah believe we not allowed to bring ganja to take a lil smoke.' Bertrand was angry that the International Cricket Council had banned some of his favourite items from the Oval. These included plastic bottles, cans, tins, large umbrellas, hard cooler boxes, alcohol and illegal drugs.

Wilfred shook his head. 'Yeah and we cyah bring we coolers and umbrellas like we accustom. Ah make sure and soak all meh sandwiches in rum and puncheon.'

Outside the Oval, a cricket fan was arguing with a policeman. The cricket enthusiast had a coffin filled with containers of peleau and sweetbread. He said, 'Boss, dis is not a cooler so I have a right to enter de Oval. Coffins is not on de list of tings banned.'

The game had already started. Marcus and the others hurried upstairs the Dos Santos Stand and located our seat numbers F058, F059 and F060. These seats were allocated to holders of season tickets. The 9th ICC Cricket World Cup in the West Indies had sixteen teams from five continents. They were competing to be cricket champions. Bermuda cricket team was the only new participant. The teams included New Zealand, Pakistan, Australia, Canada, Zimbabwe, Bangladesh and Kenya. There were numerous local and foreign cricket enthusiasts with their pocket radios keenly listening to the game which had the ingredients for a nail-biting classic.

A nuts vendor, known as Jumbo, shouted at our section of the crowd, 'Salt and fresh.' Whenever the crowd clapped or cheered he would briefly turn his attention to the game. He had a yellow and green jersey with a large picture of Malcolm X. He carried two bags. One on his shoulder and the other he held with his right hand. Yellow plastic wrapping protruded from one of the half-opened bags. 'Nuts, nuts, who want a bag of nuts?' He faced reduced sales due to the high cost of six dollars per pack which was doubled from the usual cost of three dollars. Furthermore, the presence of other nuts vendors meant more competition for Jumbo.

The delivery of the nuts to customers was entertaining and unique. Jumbo would be attracted to an upraised arm or shout among the spectators. Then he would proceed to throw one or two packs of nuts to the person. The customers would either send the money to him via other spectators or he would meet them and collect the money. A few unsuspecting persons were hit by flying packs of nuts but they did not complain and laughed at the comical situation. Some persons in the higher rows requested nuts to simply witness the throwing abilities of the nutsman.

Another vendor with a small cooler on his shoulder shouted, "Cold Carib and Stag....Carib and Stag."

Bertrand casually removed a joint of ganja from his socks and lit it. 'Dem should know we West Indians is ah free people. We go always find a way to beat the law.'

Wilfred lifted his jersey to reveal four bottles of perfume strapped to his chest. 'All dese filled with Vat 19. Imagine dey selling a bottle of beer for eighteen dollars and water cost twelve dollars a bottle!'

A patriotic citizen with a large flag of Trinidad and Tobago was circling the border of the cricket ground. Nobody knew his name or the purpose of his mission. He was known as the 'Flagman.' The Flagman had a red cap and wore a black, long pants and red jersey. He was sweating profusely and tourists were snapping photographs of him. Some members of the Trini Posse would clap and offer words of encouragement after he completed each lap.

One hour before the day's play, the groundsmen were on the cricket pitch with a small roller. Kyle occasionally peered through his binoculars to check the players' expressions. In the past, he had always wistfully observed other spectators listening to live commentary during play but could not afford a radio. Now the moment was finally here! The radio was switched on and the station band adjusted. The spectator nearby had a radio which irritated Marcus. The others were not distracted and remained glued to the game.

Marcus wished he had borrowed a radio. He glumly continued to watch the thrilling One-day match between the West Indies and Australia without the advantage of live commentary. The crowd began a Mexican wave during the water break but this fizzled out after one round. At 11am there was a slight drizzle and the groundsmen ran unto the fields with tarpaulin covers to protect the cricket pitch. Someone shouted, 'Anytime dey stick three stumps in de ground yuh sure to have rain.' Kyle had heard this remark on other occasions. He smiled. It seemed true.

Bertrand told his friends that he won free cricket tickets in February 1965 in Shell's 'Tickets for the Test Competition.' He completed the entry form which was published in the *Trinidad Guardian* and attached a receipt indicating purchase of gasoline at a Shell service station. And, his name was selected for two tickets. 'Lissen, two of the best games I ever witnessed was in June 1975 when the West Indies won the inaugural World Cup at Lord's cricket ground in England. Clive Lloyd was the captain. He made a captain's knock of one hundred and two runs. Then there was that over by Michael Holding in 1981 at Kensington Oval. Holding bowl Geoff Boycott for a duck in the first innings and in the second innings Boycott only made one run and his wicket was again taken by Holding.' We listened in awe to these snapshots of West Indian cricket by someone who had witnessed the games.

During the lunch interval, Wilfred, Bertrand and Kyle went downstairs and searched for a booth selling drinks. The ground near the bar was littered with the covers of beers and nearby there were four bins overflowing with styrofoam cups and empty beer bottles.

There was a long line at the bar. Patrons did not seem concerned with the long wait to purchase alcohol. Along the counter there were shouts of 'Five cold Carib,' 'Yuh have Red Stripe? and 'Gimme a Shandy.' The customers acted as if the louder they shouted, the faster they would be served. Wilfred bought two bottles of beer.

A young adult looking barely over eighteen years quietly made a query, 'You have any Johnnie Walker?'

'No, it jus' finish. Only Carib, Stag, Carlsberg, Mackeson, Royal Extra and Red Stripe,' replied the exhausted barman.

'Gimme a bottle of Red Stripe and three bottles of Mackeson.'

The busy barman looked at him and nodded. He did not bother to ask for any identification. The barman slid the bottles along the counter and collected the money from the customer. The barman then looked at the next person in line.

The aroma of freshly cooked, mouth-watering food filled the air. A long, curling and unbroken line began by a woman selling corn-soup. Her brisk sale was slow when compared to the vendor selling East Indian dishes in the Ali's Doubles and Roti shed. There were two vendors in this shed. One stood over a stove with two boiling pots of oil. She gently placed some rolled flour into the bubbling oil. The second vendor was collecting money and taking orders.

'Yeah. Tantie give me two more doubles with slight pepper,' shouted one teenager whose face was scarred with acne marks. His face soon became red in the sun.

The vendor neatly wrapped the doubles in wax paper and both were placed in a small brown bag. After eating each doubles, the teenager licked his fingers which had pieces of channa and sauce. Nearby were two adults who were licking the paper that had been used to wrap their doubles.

Wilfred's hoarse voice shouted from amongst outstretched hands, 'Tantie, the pepper hot? Give me a chicken roti and a solo.' He turned his head, 'Kyle and Bertrand what allyuh want?'

'A potato roti will be fine,' said Kyle. He was staring at two tourists who were wondering which of the foods did not contain pepper.

Bertrand nodded. 'Same here. Put plenty pepper.'

A hungry buyer stopped shouting for food and turned around and angrily said, 'Man doh mash meh foot na... you think meh foot is a brakes?'

A scruffy man in his early thirties who seemed a regular customer shouted to get the vendor's attention, 'Moms is me again! I want two aloo pies and a Chubby.'

'Hold on, ah coming to yuh,' replied the flustered vendor. 'Slight, medium or plenty pepper?'

The man paused. He was not sure of the answer. 'De pepper hot?'

'A little.'

'No pepper on de aloo pies. Ah have meh own pepper upstairs.' The customer seemed eager to purchase his food and return to his seat.

In an effort to avoid the chaos and long lines at the food outlets Bertrand decided to place some money on a lottery number. The line at the Play-Whe booth was long and slow-moving. He thought about which number would play in the Play-Whe draw at 1 pm. Hog? Centipede? Cattle? He saw the pastor from a nearby church and felt it would be best if he played 'Parson man.' Eventually, after considerable procrastination, he placed five dollars on 'Dead man' since the Australia would certainly be defeated in this game and considered dead in the cricket world. He thought about placing money on another mark. This one was to be associated with Brian Lara. He decided to bet more money on 'King.'

It was ironical that the number twelve was 'King' as most Trinidadians and West Indians considered Lara to be the undisputed king in cricket. He holds the record for the highest number of runs in the county cricket and Test cricket. Some cricket enthusiasts felt Sachin Tendulkar, the prolific batsman of India, was better than Lara.

A woman in tight fitting jeans was at the foot of the stairs. Someone heckled her, 'Sweet ting, come and sit down by meh na.' She ignored the heckler.

Kyle, Wilfred and Bertrand returned to their seats and observed eight empty beer bottles on the ground. Bertrand felt like returning to the Play-Whe booth and bet some money on the number eight. However the long line was a deterrent.

A red-skinned lady wearing a green jersey and short blue pants was selling small flags and commemorative brochures of the World Cup competition. A heckler teased her. 'Reds yuh looking hot in dem pants. Come and read one of dem magazines for me.'

She looked at him and smiled. 'Why yuh illiterate?'

The section of the crowd who heard the conversation began to laugh loudly. The heckler felt ashamed and remained quiet for the rest of the game.

The Australians had amassed 323 runs in their allotted fifty overs. The West Indies team had 299 for 7 and Lara and Courtney Browne were still at the crease. The West Indies team looked safe. Shane Warne, the feared Australian bowler, was becoming frustrated as Lara smugly danced forward and elegantly stroked the ball for four. The spectators on the cycle track frantically waved their plastic signs with the number four. A man with a conch shell blew loudly into it. This conch-blower was a regular fixture at the Oval and would make noise whenever a West Indian batsman scored a boundary, a six, reached a milestone or captured a wicket.

The score board read 303. Then suddenly the dream turned into a nightmare. Shaun Pollock, a fast bowler captured the prize wicket of Lara. The wicketkeeper, Mark Boucher, jubilantly jumped into the air. He caught the ball and claimed that Lara had edged it. The umpire raised his hand to signal that Lara was out. The Australian fielders rushed to congratulate Pollock and Boucher. The crowd was silent and then some booed and steupsed. Lara briefly hesitated. He stared in disbelief at the umpire and began walking to the pavilion.

Wilfred mumbled, 'Dey always tiefing Lara out…he must be holding de record for the most amount of tief out.' The slur in his voice indicated that he was no longer sober.

Kyle nodded and replied, 'Dem Australians like to cheat too much! If is not cheating is sledging. Dey know Lara is de danger man. Ever since de Prince of Port-of-Spain make 375 against England in 1994 right here, all dem teams 'fraid him.'

The drumming and shouting from the Concrete Stand Posse signalled the impending victory of the West Indies team. This atmosphere of celebration failed to change the festive mood of crowd. The West Indies needed seven runs from six balls. After three balls were bowled and only two runs scored, Shivnarine Chanderpaul was run out. The umpire signalled for the television replay to determine if the batsman had reached the crease. The crowd became anxious and heads were turned to see if a green or red light would appear on the screen. A slow motion replay would determine the fate of the unfortunate batsman. The nutsman stopped selling and also stared intently at the screen. The red light flashed and a glum Chanderpaul began walking slowly to the pavilion. He placed the bat under his arm and began removing his helmet and gloves.

It would be a nail-biting finish. Bertrand's face was wrinkled and he began to sweat profusely. Marcus and Wilfred were focused on the game. Kyle clutched the silver cross which hung around his neck. Then the diminutive West Indian batsman, Dinesh Ramdin, who was the team's wicketkeeper, hit the ball high into the air and an Australian fielder, Herschelle Gibbs, began running to catch it. However, the ball slammed into the boundary.

The ordeal was over. The West Indies team had snatched a victory from the powerful Australians. Marcus used his right hand to make the sign of the cross over his chest and wiped tears from his eyes. Most of the spectators were jumping up and down. Four women in the front row began to dance whilst holding beer bottles. Even some of the Australian tourists were happy to witness such a thrilling end.

Fans from the cycle track and Republic Bank Stand slipped past the police and sauntered on to the field. Some jumped over billboards and dashed to hug the batsmen. Three spectators began to fight over one of the stumps. They wanted it as a souvenir. Ecstatic supporters began to descend the stairs. Everyone was rushing and pushing as if the gates of the stadium would be closed in ten minutes and they would be trapped.

Marcus and the other men were exhausted with the heat and wanted to return to their homes. They decided to take a taxi. During the return trip, Marcus and Bertrand recalled the glory days of West Indian cricket and both men had seen the bowling and batting prowess of such cricket icons as Alvin Kallicharran, Rohan Kanhai, Desmond Haynes and Gordon Greenidge.

When Marcus and Kyle entered their home, they closed the windows and bolted the doors. Marcus said, 'Son I don't want you to be a footballer, cricketer or athlete. You can play these sports or run marathons as recreation but don't make it your job. I want you to be educated. Some of these sportsmen have a short shelf life. You need to get a university degree and get a good job. Education is de only way out of poverty and hardship.' Kyle listened to his father's advice.

Next morning, Kyle awoke and looked at the 2006 calendar on the faded cupboard. He did not want to spend another year in Trinidad. He checked the internet for Canadian and British universities and decided to apply. He eventually accepted the offer to pursue a degree in Sociology at the University of Guelph in Ontario, Canada.

During the journey to the airport Kyle received advice from his father.

Marcus said, 'In the West Indies and odder developing countries, your educational achievements are an important indicator of your success.'

'Is not so in Canada and de USA?' asked Kyle. He looked at the flora and began to regret that he was leaving his homeland.

'If yuh sneeze too hard some of dem universities abroad will give yuh a scholarship and credits towards yuh degree. Ent the brochures yuh show me giving credits for extracurricular activities?'

Kyle quickly replied, 'Yeah.' He smiled and enjoyed the cynicism.

'Well dis is exactly de ting ah trying to say.' Marcus used illustrations to emphasise his point. 'If you duncey buh could sweep de classrooms or play football in de park den de university go give yuh credits. Doh go abroad and geh fooled dat yuh bright and educated. Do yuh extra reading and learn to improve yuh writing. Yuh would be surprised to read some of de bad spelling and grammar from people at North American universities. And den it have affirmative action. If yuh is a minority group yuh getting a place at the university and a scholarship. And, please...do anyting buh doh bring back a white gyul. She will say she love you but soon leave you. Anyway do I have to warn yuh 'bout bad company?'

'No.' It was concise and blunt but advice that he appreciated. Kyle smiled. He understood when his father was being sarcastic and providing wise counselling.

'Good, well have a safe flight.' Marcus helped Kyle carry his two heavy suitcases to the end of the line at the counter of Caribbean Airlines. Kyle slowly walked to the check-in counter and then to the departure lounge. As he boarded the plane he felt leaving Trinidad was a bad decision.

After six hours of flying, the airplane arrived in Toronto. There was some turbulence but the flight was an enjoyable one. Upon clearing the customs, Kyle took a bus to Guelph and went to the Student Administration. After registering and collecting documents, he settled into his room. The academic journey had begun.

Guelph is a bustling town with a relatively small population of 100,000 persons. The Ontario Veterinary College and the Ontario Agricultural College on the campus attracted students from around the world.

Kyle's friends were surprised that he chose to study Sociology at the University of Guelph because this was an institution noted for its pioneering work in fields such as animal and plant biotechnology, toxicology and child development. He was expected to gravitate towards the University of Toronto or York University. His decision to be in Guelph isolated him from the majority of West Indian students, at Toronto, and allowed him to have a genuine Canadian experience.

His stay in Guelph and visits to nearby communities proved to be most memorable particularly because it was his first extended stay in a foreign country. The decision to rent off-campus allowed him to experience student life and interact within a community.

The educational environment in Canada was markedly different from Trinidad. The university students at Guelph were more radical, environmentally-conscious and vocal on topics such as globalisation and neo-imperialism. This was evident in the various clubs and societies which dealt with a host of religious, social and political issues. It seemed strange that students would spend hours, fervently discussing ideas in the bistros, cafes and on the campus grounds. There was no apparent dichotomy between academia and political or social activism. Despite this interest in international affairs, the students and residents of Guelph had little interest in the provincial or national politics. This probably stemmed from the fact that their town lacked serious social problems such as crime, poverty and unemployment. Interestingly, there was one television station with live debates from the politicians at the Parliament based in Ottawa. Politics in Canada seemed civil when compared to politics in Trinidad and Tobago. There seemed to be an absence of the racial dimension and picong which were distinct features of the politics he was accustomed to in Trinidad and Tobago.

In Trinidad, these social issues did not motivate the students to adopt drastic action. The university students in the Caribbean were more concerned with meeting academic deadlines or enjoying weekly parties. Undoubtedly Caribbean students tended to believe that social and political issues should be confined to debating clubs or dealt with by flamboyant politicians and older adults.

At Guelph, the university's newspaper, *The Ontarion*, regularly highlighted the injustices meted out against developing countries. For instance, the newspaper clamoured for the boycott of international companies which employed child labour in the manufacturing of their products. Kyle wondered if Canadian politicians cared or knew about these injustices. The writers were outspoken and boldly questioned the status quo. The newspaper was free of charge and provided a vital educational tool for those students who were not inclined to read much or could not afford the local *Guelph Mercury* or the major provincial newspapers as the *Toronto Star* and *Globe and Mail*. *The Ontarion* was circulated among businesses which paid for advertisements, and served not only as the voice of the students but informed the rest of the town of the ideas emanating from its prominent university. Kyle regularly contributed one-panel comic strips which were accepted and published in *The Ontarion*. This made him more inclined to collect a copy and browse through the articles.

Kyle checked the internet for the latest results of the games in the French and Dutch Leagues. Lyon had defeated Nice in yesterday's match. He wanted Bordeaux to defeat Toulouse in the French League. Marseille was third in the standings. In the Dutch League, AZ was at the top whilst NAC and Ajax had a crucial encounter tomorrow.

He checked the Spanish League and saw that Barcelona was still at the top with thirty points. This was no surprise. The team had a consistent track record. He favoured Real Madrid which was second on the table. Other teams as Roma, Torino, Reggina and Bologna were lower in the standings.

Kyle received an email from Carla. She informed him that his grandfather's aunt, Barbara, had died. She was buried in a plot next to her two toes and her husband's leg. Kyle was relieved that she had died. His grandfather always felt the need to attend the funerals and memorial services held by Barbara and George. If he missed a service, Barbara would call and make him feel guilty. The hospital decided to establish the 'Barbara Gopaulsingh Ward' because she spent most of her life in that ward.

Six weeks later Juanita died. The funeral was held in a nearby Roman Catholic Church. Her daughters and husband did not attend the funeral. Marcus delivered the eulogy.

At the University of Guelph there were regular and vocal student protests over the proposed increase in tuition fees. This eventually culminated in a group of students, some comprising members of the Student Council, deciding to barricade themselves in one of the university's administrative offices. His position as the International Student representative on the Student Council did not jeopardise his status in Canada but allowed him to lend verbal support to the campaign. Kyle was impressed by these students' commitment to a cause and belief in their actions as being justified.

It was part of his cultural shock to see the existence of a Queer Equality club which embraced different sexualities. Kyle said, 'This is amazing that such a club could exist.'

Ellen nodded. 'The club provides a safe and supportive environment for the transgendered, bisexual, lesbian, and gay students. It would be an even greater shock for you to learn that such societies were not exceptions but rather the norm among other universities in Canada which also had similar groups and boasted of a vibrant membership.'

In the Caribbean, gay relationships are publicly condemned and stigmatised. Same sex unions are rejected by West Indian societies as abnormal, immoral and an affront to religious beliefs and ethics.

Furthermore, Kyle was surprised that the University of Guelph allowed the formation and existence of a Pagan Society on campus. But it was regarded as an expression of religious freedom. The functioning of other religious clubs such as the Inter-Varsity Christian Fellowship, Ismaili Muslim Students' Association, Muslim Students' Association and the Jewish Students' Organisation reflected the diversity of the campus. Interestingly, despite the various religions there were no synagogues, temples or mosques in Guelph. Questions bombarded his mind - how could an institution of learning encourage and perpetuate moral decay? Why did the authorities from the province or government not intervene? What was the response of the Christians and politicians to the promotion of this aberration?

Kyle approached some fellow students for an explanation of this dilemma and contradiction. 'Is being gay tolerated by the public?'

Marc, a doctoral student, sought to explain. 'Canada has tried not to discriminate against persons of different sexual persuasions or religious beliefs. Other Canadians wondered about the prudish nature of the Caribbean society. These persons did not have a psychological problem but should be given equal opportunities as handicapped persons, non-Christians and persons of colour.'

'Whatever is in the Bible, is what I follow. Nothing more, nothing less,' said Kyle. Initially, this philosophy seemed unrealistic and as an attempt to create a utopia. He learnt an invaluable lesson from the existence of such clubs - that tolerance does not mean acceptance or justification. Tolerating persons with differences means acknowledging multiculturalism and diversity but more importantly included the treatment of others with dignity and respect.

Marc spoke for five minutes then checked his wristwatch. He decided to leave.

The boundaries of the campus could not be easily defined. There was considerable fertile land under cultivation, farms and lands leased to companies. Kyle passed buildings such as the Stone Road Mall and heard that this used to be under cultivation and belonged to the university. It was easy to become a bit disoriented and lose a sense of direction in such a vast campus. During Kyle's residence in Guelph, he would frequently hear the phrase, 'All that is the university land.'

The university's architecture was unique. Initially the buildings seemed grotesque and overbearing with the landscaped plazas, modern concrete, glass and steel, Victorian turrets, hand-hewn century limestone and brick walkways. Colourful brochures of the University of Guelph which Kyle had previously internalised did not fully reflect the architecture. The brochures provided glimpses of the campus- the War Memorial Hall and Johnston Hall. However, Kyle realised that these structures symbolised revolutionary engineering ideas. The placement of buildings offered easy access during cold snowing winters and hot summers. It was most peculiar in this quaint town to observe these modern structures juxtaposed to outdated buildings.

Kyle was overwhelmed by the grandeur of buildings in nearby Toronto. Huge skyscrapers dotted the skyline of this city. A massive stadium in the heart of Toronto would cater for avid baseball and football fans. Kyle had never seen a stadium of this magnitude and was certain it could comfortably hold the population of Cunupia, in central Trinidad. He wondered about the cost of these architectural masterpieces and the date in which the stadium was built.

'Excuse marm. Do you know when this stadium was built?'

The lady stopped. 'No.' She shook her head and continued walking.

Nobody seemed to know or care about the date or cost of buildings in Toronto.

Kyle attended his first hockey game in October. He enjoyed it because it was a fast-paced game. All the players were females. He looked at the joy of the players and supporters when the puck was at the back of the opposing team's net. He had witnessed this jubiliation during football games in Trinidad. The following day he attended an ice hockey tournament and witnessed six hours of various teams competing for first place. He met Ellen and her boyfriend, Sebastien, at this game. Sebastien was from Brazil and pursuing a degree in Art at the University of Waterloo.

'Hey Ellen, how are you?'

Ellen was surprised to see him. 'Didn't realize you liked hockey. Come and sit with us.' She introduced him to Sebastien. In the summer she and Sebastien planned to visit Italy and France.

'It's my first game. They play it in Trinidad but not on ice.' Kyle compared the games of hockey and football. Both were fast-moving games and had goalkeepers.

Kyle also mentioned the increasing presence of women football in the public domain. 'There is a Women's European Championships which attract a lot of attention.'

'Really? I never heard of it.' Her eyes remained fixed on the game.

'Yes. I have not followed it recently but Norway won the title in 1987 and 1993. Countries as Finland, England, Sweden, Italy and France participate.'

Sebastien asked, 'Do you follow the Copa America?'

'I don't usually follow the various group games but watch the finals.' One of the players received the puck and hit it past the goalie and into the back of the net. 'Some countries tend to dominate the finals in the Copa America. For instance in 2007 the finals had Brazil and Argentina. If you check the winners for the past twenty years it was those two countries along with Uruguay and Columbia.'

'Yeah it's true.' Sebastien folded his arms. 'I never thought about it. It would be good to see other countries as Venezuela, Mexico or Ecuador being the champion.'

Kyle was glad that Sebastien agreed. 'One good thing about the Copa America is that certain players get good public exposure and become stars.'

'It's true. After the competition I'm more familiar with names like Maldonado of Venezuela, Villanueva of Chile, Mendez of Ecuador and Morales of Mexico.'

The game ended and Kyle bid goodbye to Ellen and Sebastien.

Kyle admired certain streets in Guelph, Elmira and Elora with flowerpots hanging from the streetlights. It was an aesthetic asset as it gave the neighbourhood a very welcome and informal atmosphere. It was the first time in his life that Kyle had seen homes without fences and burglar alarms and windows without burglar bars. It was comforting to walk past homes and not have to be cautious of noisy and vicious barking dogs. There was nothing to disrupt the tranquility of evenings. Indeed, it seemed that crime was non-existent in this section of the world. Even though there were no slums, certain buildings in downtown Guelph were neglected with unkempt lawns and mossy walls.

During the first week of September, Kyle attended a football game at the Bank of Montreal Field in the Exhibition Place which was located near the lakeshore in Toronto. At the game he learnt that most Canadians referred to football as 'soccer'. The two teams- FC Edmonton and Toronto FC, were competing in the Nutrilite Canadian Championship for the Voyageurs Cup. The diverse crowd included West Indians and Africans.

The Canada geese, moose, beaver and the maple leaf were symbols of Canadian identity and heritage. It was most peculiar to observe the pervasive patriotism among Canadians. Partial proof of this nationalistic attitude was the Canadian flag proudly fluttering on the flagpoles on front lawns.

The inhabitants in Guelph neither viewed themselves as residents of a town nor the province of Ontario but on a national level as Canadians. This absence of parochial attitudes and fondness for their country were closely protected and clearly reflected in the cleanliness of the town and the refusal of citizens to discard rubbish in rivers and roads. This mentality was glaringly absent among some young West Indians, who seemed uninterested in fostering a sense of patriotism and were extremely eager to migrate to developed countries. Secondly, most Trinidadians were devoid of sensitivity towards the environment.

One of his lurking fears was racial discrimination in Canada, especially after reading reports of the anti-social activities of skinheads who comprised a scattered group of neo-Nazis. Kyle wondered if the Whites would be receptive to coloured persons. Fortunately, there were no unsavoury incidents. During walks across campus it dawned upon him that the University of Guelph catered for an annual influx of international students. The diversity on campus was evident with international students from China, Australia and Africa. Despite being the only coloured student in the postgraduate programme of the History Department, his peers were cordial and helpful. His professors were most encouraging. Even those professors who did not teach him were courteous.

The International Student Advisor was extremely cordial and student-friendly. Kyle felt little affinity with students from developing countries. Some were physiologically similar but their culture was markedly different.

Kyle was of slim built, weighing 100 pounds and 5 feet, 2 inches in height. Indeed, this sharply contrasted with the average adult Canadian males and females who was stronger, taller, and had a much bigger physique. Even the strapped high school teenagers loomed over his diminutive frame.

A fellow student jokingly asked, 'Are you a dwarf? Have you recently graduated from primary school?'

Kyle laughed. He ignored such questions and felt some consolation when he was among the Chinese and other Asian students who were similar in physique. Kyle met Jing Yang of China who was an avid football fan. He was a professor in the Chemistry Department.

'So you are from Trinidad and Tobago. Did you go to Europe to support your team in the World Cup a few years ago?' asked Jing. He sat in his office which had been his second home for eleven years.

Kyle had a sad expression on his face. 'No. Could not afford it but saw all de games on tv. I sometimes see highlights of games in de Asian Cup. De standard is very high. Did you see the exciting Asian Cup game between South Korea and Iran which South Korea won on penalties and advanced into the semifinals?'

'Yes. Javad Nekounam almost scored for Iran when his right-foot shot touched the goalpost and also Mehdi Mahdavikia blasted a long-range strike which the Korean keeper saved. I was glad to see the other games where striker Yasser Al Qahtani scored one goal and helped create another to ensure Saudi Arabia beat Uzbekistan 2-1. All the foreign players get a lot of money to play in the games.' Jing occasionally played football with other academic colleagues in the park. 'Games in de Gulf Cup are also exciting. Were you surprised when Al Wahda won the United Arab Emirates League title?'

'Yes, dey have some good players. I thought Al Jazira was going to win. Have you realized dat de world of football has changed?' asked Kyle.

Jing looked puzzled. 'How?'

'In the twentieth century, the favourites were Argentina, Germany, Italy, France, Brazil and England. Dey dominated football. One of dese teams was expected to be in de World Cup or were predicted as World Cup winners.' Kyle paused. He wanted to hear the response of Jing.

'It's true what you say. The media and most of the spectators focus on a few personalities as a result of goals scored in qualifying games, international matches or club games. But there are a lot of good players who go unnoticed.' Jing scratched his head. 'For instance, in previous World Cups there were good defenders as Naybet of Morocco and Mihajlovic of Yugoslavia. There was also Radebe of South Africa, Prodan and Popescu of Romania and Guardiola Sala of Spain.'

'Exactly what I was saying. Not many people would recall the names of talented midfielders as Beya of Tunisia and Gilles de Bilde of Belgium. Dese men did not get public recognition because dey were not playing on the winning team or a big team like Brazil or Italy.' Kyle looked at the students in the line waiting to purchase coffee. 'Sometimes there is a colourful personality who briefly captures the public's attention. Remember Valderrama of Colombia?'

'Yes, yes the midfielder with the big and loose orange hair,' replied Jing. He smiled. 'I saw him play in the 1990 and 1994 World Cups.'

'I have a strange feeling that an underdog will win the World Cup in 2010.'

Jing smiled. 'Me too. Do you know that there is an international tournament known as the Homeless World Cup?'

Kyle's eyes brightened. 'Yes, I heard about it but don't follow the matches. I think it is an excellent idea to allow the homeless to perform in a world-class game. It began in 2005 or 2006?'

'I believe the first was in 2003 and held in Austria. One of the objectives of the tournament is to overcome homelessness. In 2011 it is in Paris and I will attend and also visit my sister. The rest of the world needs to deal more seriously with these issues like unemployment, poverty and vagrancy.' Jing checked his watch. His class would begin in ten minutes. 'Okay, my class begins soon. Nice talking to you and we will meet again. What did you say your name was?'

'It's Kyle. Nice meeting you too. Bye.' He exited the building and headed to the library.

One of the indicators in assessing Kyle's response to Canadian society was his adaptation to the winter conditions. He monitored, on the television, the fluctuating temperatures, wind chill and ultraviolet radiation. Detecting and avoiding patches of ice in the snow gradually developed into a skill. One friend told him that only when he fell on the ice could he truly appreciate the winter.

Kyle had undergone a metamorphosis with the layers of warm clothing, winter shoes, woollen cap, scarf and chapstick to protect his lips from the cold. The thick layers of clothing made him look as if he had suddenly gained four or five pounds. During winter it would be difficult for anyone to judge that he was from the Caribbean. Furthermore, his diet was slightly modified as it included hot soups, coffee and tea to keep his body warm.

Adjusting to the food in Canada was part of the culture shock. Kyle did his weekly shopping at the nearby No Frills grocery. His diet comprised mainly of 'channa' which was a legume, and usually sold in packs and roti which was a type of flat, circular bread. The Canadians referred to 'channa' by its biological name- 'chick peas' and their concept of roti was pita bread which was thicker and different in taste though. Kyle was accustomed buying items which were uncooked and spending time cooking and adding various seasonings. In Canada, an overwhelming majority of the foods were pre-packaged and ready-to-eat.

Canned soups, frozen television dinners and potato chips, simply needed a few minutes of heating to be edible. It was obvious that a microwave was essential in any kitchen. For many university students, convenience rather than health was the major concern. Reducing the time spent in the preparation of meals was welcomed by busy households. However, the consumer was deprived of natural and fresh foods and also the feeling of achievement and originality in preparing a meal.

Sampling the new foods was an absolute joy. Kyle attended the annual Maple Syrup Festival in Elmira. He was thrilled to finally taste the maple syrup. It was a delicious substitute for honey which was used on pancakes. Similarly, his visit to the Kitchener Farmer's Market was rewarding. He sampled the apple butter, bacons, schnitzel and local cheeses. It was the first time he had seen Mennonites who were easy to recognize with their distinctive clothing and hats. The Oktoberfest, the Bavarian festival, in Kitchener-Waterloo was amazing. To be among the thousands of spectators witnessing the clowns, floats and bands was a memorable experience. The atmosphere was similar to Carnival in Trinidad but without the scantily clad revellers and wining.

One noticeable feature of Guelph and the surrounding communities was the Bed and Breakfast dwellings. Each community had first-class accommodation was like a mini five-star hotel. It was something that could not be easily overlooked by a tourist. Guelph had the Willow Manor. And, in Elmira there was The Evergreens, The Foreman's House, Teddy Bear Bed and Breakfast Inn and Blossom Hill. Among others were Blair's Little Acre, Cressman Guest House and the Dappled Pegasus Farm in the Kitchener-Waterloo area. These places were located in relaxing and scenic surroundings. They served tasty breakfasts and dainty furniture. In Trinidad, Kyle was more familiar with the motels and guest houses which served as drug dens or strip joints. Some such as Rich Gold, in Chase Village, in central Trinidad, were infamous as meeting places for illicit liaisons and often employed illegal migrants.

It is difficult to overlook the plethora of museums, archives and heritage sites scattered across the province. He visited a few in the nearby towns of Waterloo, Elora, Fergus and Niagara. These included Uncle Joe's Elora Antique Market, the Seagram Museum and Canadian Clay and Glass Gallery in Waterloo and the Wellington Galleries. He did some research on early Black Canadians at the Wellington County Museum and Archives. He was impressed with the annual Festival Theatre in Stratford which showcases quality plays such as King Lear, The Music Man, Amadeus and The Merchant of Venice. The existence of such places was the most potent symbols of how appreciative Canadians were of their past and culture.

Kyle entered the library and checked his email. After he read the results of the games played in the Primera División in the Mexican professional football league. The Clausura tournament was held in the summer and the Apertura tournament occurred during the winter months. There were eighteen clubs competing in these tournaments. Kyle was impressed with Pumas which had won six titles.

The province of Ontario possessed extensive Canadian art and sculpture galleries. An appreciation for the arts is obvious in the MacDonald Stewart Art Centre on the campus of the University of Guelph and the Kitchener-Waterloo Art Gallery. Furthermore, the vibrant Ontario Heritage Foundation had a newsletter—*Heritage Matters* which Kyle regularly read. He was able to identify with this desire to carefully preserve for posterity. In contrast, Trinidad and Tobago was a philistine society which failed to appreciate the significance of preserving the past.

Conservation and environmental awareness comprised an integral component of Canadian life. The parks and children's playgrounds were tidy with such amenities as picnic benches and toilets. One of the dynamic organisations on campus was the Wildlife Club which increased the awareness of maintaining an ecological balance and preserving the environment. Guelph embarked on an impressive recycling project which was implemented in all neighbourhoods. Blue and green plastic bags were judiciously utilised to separate household refuse into dry (bottles and cans) and wet (food leftovers) items. It was fascinating to observe communities in Guelph cooperating to ensure the successful implementation of this venture. The town had a premier recycling plant which complemented the efforts of this progressive program.

On Sundays, Kyle attended either one of two nearby churches—Arkell United Church or the Trinity United Church. It was a surprise to realise that the congregations in both churches comprised mostly elderly persons and retirees. He was taken aback by the glaring absence of young adults and teenagers. This was probably an indicator of a lack of interest in the form of worship of traditional Christian denominations as Methodists, Baptists and Presbyterians. This trend also suggested less reliance on religion as a source of comfort during times of distress or for faith in God.

His impression of Guelph was that there were no homeless persons. This might have been a flawed judgment but he did not encounter any beggars or street children. He had seen homeless persons in Toronto. It was odd to see vagrants who were light-skinned. In Trinidad and Tobago, and the rest of the Caribbean, he had been accustomed seeing Afro-Trinidadian and Indo-Trinidadian beggars and street children. The other ethnic groups- Whites, Syrians Portuguese and Chinese were usually the elite and wealthy persons in the Caribbean. They were successful business persons, owners of commercial buildings and malls. Their funding often determined the success or failure of a political party.

Undoubtedly, an overwhelming number of photographs in magazines and documentaries tended to focus on poverty and social problems as crime among coloured persons in developing countries or in the slums of First World countries. Thus, Kyle always associated the curse of poverty with coloured persons.

In Guelph and Toronto, poverty seemed to be eradicated or reduced to a minimum, and the middle class residential houses in the suburbs attested to the fact that unemployment was low. It was difficult to grasp the fact that all Canadian citizens were covered by health insurance which allowed them to be treated in hospitals which were clean, efficient and well-equipped. This was a surprise, since in Trinidad only those who could afford to pay for medical insurance were guaranteed proper health care in the private nursing homes and hospitals. In many Caribbean countries, the poor who were sick had no alternative but to endure the shabby public hospitals.

Whilst he frequently made comparisons between Trinidad and Guelph in terms of education, the efficiency of social services and nationalism, Kyle was also acutely aware of Trinidad and Tobago's disadvantaged status as a developing country. Secondly, Trinidad had certain desirable advantages, such as a tropical climate and beaches, which Guelph did not possess.

Kyle was fortunate to be interacting with persons who were understanding and willing to explain the various social mechanisms which promoted coherence in the fabric of the Canadian society. The warmth and friendliness of the Whites in the neighbourhood, campus and congregations was the most positive aspect of his brief sojourn in Canada. Kyle wondered if the cordial reception of the Canadians stemmed from their belief that he was a student, a temporary visitor, who would soon return to his homeland and not be a burden on their social services and compete on the job market with other Canadians.

Chapter 8
False victories

In April and November 2009, the Government of Trinidad and Tobago decided to host two international conferences- the Summit of the Americas and the Commonwealth Heads of Government Meeting. The conferences were held in the country's capital, Port-of-Spain. It was an appropriate venue because Trinidad and Tobago was the talk shop capital of the world. The citizens loved to talk. They talked in rum shops, on the streets, in television and radio programmes, in taxis and buses. The talking heads continued even if nobody was listening. The public grew accustomed to an expensive Commission of Enquiry in which nobody would be convicted. The citizens talked of solving local problems and the actions to be taken if they were prime minister. Unfortunately, the talking and grand ideas never materialized into solutions.

In preparation for the conference, the government decided to build a high wall along the highway to hide the squalor and poverty of Beetham Gardens. In March 2009 the government agreed that the vagrants and homeless persons in Port-of-Spain were an eyesore and felt they should be placed in the St. Ann's Psychiatric Hospital. However, there were no extra rooms at the Psychiatric Hospital for the street-dwellers who were also referred to as vagrants and homeless persons. To create rooms for the vagrants and homeless, the government decided that the majority of patients at the Psychiatric Hospital would be discharged and returned to their families or sent to another part of the island. Government officials organised a lavish graduation ceremony for patients of the Psychiatric Hospital. The ceremony was held at the Hilton Hotel and each patient was given a certificate of sanity. Trophies, plaques and medals were given to patients who were well-behaved, clean and obedient.

Two days after the graduation, the Port-of-Spain City Council was given the onerous task of capturing the homeless and vagrants and transporting them to the Psychiatric Hospital. Aboud was one of the employees of the City Council who was responsible for ridding the streets of undesirable persons. Most of the beggars and vagrants read and heard of their impending fate. A few of the street dwellers decided to temporarily move into nearby communities of Belmont and Beetham and then return to the city when the conference ended. Raphael was one of the vagrants who had been captured. He suffered from dementia and usually spent a few days at the Centre for Socially Displaced Persons located near the East Dry River. Residents of the Centre received food prepared by the nearby St.Vincent de Paul Society. Raphael did not like the other residents of the Centre. He wanted freedom and decided to leave. He was sleeping on a park bench when he was rudely awakened by employees of the City Council. They roughly dragged him to a bus and he enjoyed the ride to the Psychiatric Hospital. He was excited to be in his new home and thought it was a hotel.

After the capital was cleared of all homeless persons, the Port-of-Spain City Council decided to capture stray dogs and cats. Aboud found that catching the stray animals was much easier than chasing after the vagrants. The exercise of capturing vagrants and stray animals was repeated in October 2009. The foreign delegates were impressed that Port-of-Spain was clean and scenic. The government of Trinidad and Tobago proudly announced that they had eliminated poverty and would soon tackle unemployment.

In April, Obama arrived in Trinidad. This was a memorable moment. Wilfred, former owner of a funeral home and now a local television reporter, was in awe. He told viewers, 'President Obama is coming down de plane all by himself. He took nine seconds to descend de stairs. Look, he is smiling a lot. De President can smile well. He is waving to the crowds at the airport. This is unbelievable. Look, he is now entering his car.' His immature coverage of the arrival of Obama reflected the low level of journalism that was fed daily to the citizens of Trinidad and Tobago.

Hundreds of residents of Beetham Gardens sat on the infamous wall, built by the government, to watch President Barack Obama and his entourage comprising more than thirty vehicles. Among those seated on the wall were Mark, Jameel, Neveal, Syl and Derick. The elderly men felt proud to be residents of Beetham Gardens and being able to witness this historic moment. Mark suffered from high blood pressure and Syl was partly paralyzed due to a stroke. The poor and unemployed were delighted to view the vehicle carrying the first Black president of the United States. Obama was seated in the backseat of the heavily tinted 'Beast 1' and on the car's hood were two small flags of United States and Trinidad and Tobago. Scores of soldiers and policemen patrolled the Beetham Highway to ensure the residents did not try to stop the president's entourage.

Whilst the majority of residents sought to catch a glimpse of Obama's car passing on the highway, seven boys were enjoying a game of street football in Beetham Gardens. These boys neither cared nor knew of the whereabouts of Obama. They did not care to see the famous president and he could not help them.

Carla worked as a receptionist at the Hyatt Regency Hotel in Port-of-Spain. She was able to see world leaders including Obama who stayed at the hotel for the Summit of the Americas. She laughed when the Trinidad and Tobago's Foreign Affairs Minister broke protocol and hugged a departing

Obama at the airport. She sent a text message to her father, 'd minister lucky de secret service did not put a bullet in she head. LOL. who she feel she is 2 hug de prez. she think he is rock star or singer, laterz.' She sent a similar note on Twitter.

In November, hundreds of residents of Beetham Gardens again sat on the infamous wall and waited for the vehicles transporting the Queen of England and other world leaders. A few residents believed the government had built the wall so they could have a comfortable seat to view the dignitaries who passed on the highway. They were not bothered about the millions of dollars wasted on the conferences. They would still be supporting the ruling political party at the polls.

In December 2009, Neveal and Derick died of prostate cancer. It was also the month that stray animals and street-dwellers returned to Port-of-Spain. Life would continue as normal. For the homeless their return to the city was a victory. The majority of citizens soon realized too much money was wasted at these conferences and felt there was need for a change in the political regime. Disgruntled persons were eager for an early election to vote against the government.

Malcolm awoke at ten o'clock. The sky was overcast in Trinidad. He went to the kitchen and ate a banana. He could not find any muffins and decided to cut two slices of watermelon and placed them on a plastic plate. He opened a small pack of yogurt and began reading the ingredients. It had preservatives, colouring, sugar and dyes. He was hungry and ate two packs.

Last night he attended the annual 45 Shop Lock competition. The host was Jugglers and the participants included Detrimental and Rapid Response, Delegates and Coppershot. Tonight he planned to attend a concert featuring Bounti Killa, Shaggy, Coco Tea, Movado, Kartel, and Capelton. Next week he and his friends would travel to Port-of-Spain to attend a concert featuring the sons of Bob Marley.

He sat on the couch and switched television channels. He decided to watch the rerun of an election meeting on CNC3. It was a meeting of the People Politely Partying, the opposition political party, held at Skinner Park in San Fernando. A young Indian artiste was singing 'Go na, George leave and go na.'

Malcolm sang along. He was a supporter of People Politely Partying and hoped they would win the elections. He wore an orange jersey with the words- 'Imelda for Change.'

One of the speakers shouted in the microphone, 'George tell de people where de money gone? Yuh waste billions of dollars. When we party come into power we will serve the people, serve the people, serve the people.'

Another speaker slapped his chest and shouted, 'We is de government.'

The huge crowd at Skinner Park enjoyed the rhetoric and slogans. Many in the crowd and television viewers did not expect most of the party's promises to materialize. The sun was hidden among clouds. There was a drizzle but no threat of a downpour.

There was a loud knock at the door.

Someone shouted, 'Good morning. Anyone home?'

Malcolm steupsed. 'Oh damn, ah fed up tell all yuh people to leave meh alone on a Saturday morning. Ah is ah Methodist. Leave de *Awake* and *Watchtower* magazines in de damn postbox.'

'Sir we aint no Jehovah Witness.'

Malcolm peeped through the curtain. There was a group of persons wearing yellow jerseys and orange caps. He recognized three political personalities. 'WAIT. All yuh from de People Politely Partying? Ah coming now, now.'

Malcolm was bareback and went searching for a clean jersey. He shouted, 'Justin geh yuh lazy arse off de bed and come and meet Devon, Carol, Imelda and Clyde. Harriet wake up and come and meet members of de Power to the People Party. Yuh always want to meet dem!'

There was a muffled response from the bedroom. Someone was coughing. Malcolm went to the entrance of the bedroom and shouted, 'If yuh miss dem now yuh might have to wait five years.'

Imelda, Carol and Clyde had embarrassed expressions.

Devon shook Malcolm's hand. 'Hello young man. We are members of the political party- Power to the People Party. We will save Trinidad and Tobago from bad governance.'

Malcolm was concerned with reports he was hearing and wanted to get a better perspective of the elections. 'Yuh tink yuh could win de election with de present government propaganda machine working so good?'

'Yeah,' replied Devon. He was confident. 'De people have reach ah limit and cannot take on all de corruption in de building projects. De people against smelters in south and de steel mill in Claxton Bay.'

'Buh de government is a solid force. They vote for partee and not leader. Some of dem supporters of de government doh care if dey get cancer from the smelters and steel mill. De government saying dat if anybody get cancer or tumours dey will be treated well in de first class hospitals in dis country such as San Fernando General.'

Clyde smiled. He adjusted his red beret and red spectacles. 'Most of de people who vote for de government now voting for us to get rid of George.

Some will not vote but de strategy on the ground is to vote against George. Look at their meetings…see how many government buses it have and is de same people at each meeting.'

'De people…wants to geh rid of Emperor George,' said Carol.

Malcolm shook his head. 'So what is de plans of yuh party to reduce unemployment and crime? When de larse government was in power they wasted millions of dollars to host the Miss Universe competition. Dat money could have been used to help poor people and buy medicines for de hospitals. De government den say dat having de beauty competition will help our country with trade and investment. Suppose dis continue if allyuh win de election?'

'We will rise higher than them,' responded Imelda.

Clyde handed him a booklet with colour photographs. 'Here read dis. Is we manifesto. We are going to build a seven billion dollar highway.'

Malcolm said, 'Ah find all yuh looking very different in dis booklet and de newspaper ads.'

'How?' asked Carol.

'Well all yuh teeth straight and yuh skin fairer in de pictures.'

Clyde replied. 'Yes, everyone in de Power to the People Party is charismatic and photogenic.'

Malcolm said, 'De money for de highways could be better used.' He pointed to the door, 'Why allyuh not fix de existing roads like dis one outside? Why not repair existing schools and hospitals? Wouldn't dat be more cost effective dan building new ones? Why not put more medicines in de hospitals?'

Clyde smiled. 'If we get ah loan from IMF, IADB or World Bank is to build new tings not repair.' Carol and Imelda were serious.

Malcolm was unhappy with the reply. 'What 'bout de President's House dat collapse? Yuh planning to repair dat?'

Carol decided to answer. 'Yes, that is a good question. That is a different story. Once we get into power, we will immediately start work on repairing it.'

Justin emerged from a bedroom. 'Morning everyone.' He was wearing a jersey with a symbol of two crossed arms and under the symbol had the words- Do So.

'Good morning,' said Carol. 'Come and join our walk.'

Imelda had a broad smile. 'Good morning. We will rise. How old are you young man?'

'Morning. I'm fifteen years old.'

She bit her lower lip and then asked Justin, 'Do you know that if we win the next electon you will get a free laptop?'

Justin remained quiet for a few moments. 'Yes. Buh ah already have a laptop.'

Imelda smiled and patted his head. 'We will rise!'

Harriet emerged from the bathroom. Her eyes were puffy and she wore an old yellow dress that needed ironing.

Malcolm introduced Harriet to the visitors. 'Dis is Harriet meh sister.'

Harriet was embarrassed and quickly brushed her hair and placed it in a ponytail. She was employed as a security guard. 'Sorry for meh condition. I'm working on a late night shift for the past two months. Sorry 'bout wearing dis dress, all we good clothes washing and ah have to buy more blue soap from Allum's Supermarket.' She shook the hands of the visiting group and then sat on a couch near to Clyde.

'Dat is okay,' said Devon. He removed his cap and wiped his oily head. 'Remember crowds don't win elections. We need you to come out and vote on election day.'

'We always follow de political meetings on de radio,' said Harriet. 'And for many years we going Price Smart, Centre of Excellence and Trincity Mall for meetings.'

'So wot services does your community need?' asked Clyde. He began distributing a pamphlet.

'Me? Yuh asking me?' asked Harriet. She took the pamphlet. She was shocked and never believed her opinion mattered. 'Well we want better roads, running water and de crime real bad here. For donkey years we squatters suffering.'

Imelda responded, 'We will rise!' She adjusted her broad-rimmed hat and checked her wristwatch.

Justin nodded and began biting his fingernails. Harriet scolded Justin, 'Boy, doh shake yuh head and take yuh fingers out of yuh mouth. Yuh in de presence of good politicians.'

Imelda smiled. 'We will rise.' She looked at the shelves with ornaments in the small living room. It was a humble family home.

Devon asked, 'Young man, are you still in school?'

Justin bent his head. 'No sir. I use to attend Pleasantville but ah leave school larse year and work for three months in Subway in Gulf City Mall and one month as a cashier in Hi-Lo gorcery. On weekends ah does sell pirated movies and songs on High Street in San Fernando.'

Harriet poked him. 'For Chrissake speak proper English. And raise yuh head when yuh speaking to people. Sir, life is hard for poor people in Marabella.'

Devon nodded. He began dialling on his cellphone. He said, 'Hello is Dev, tomorrow send some workmen on de Line in Marabella.' He closed the cellphone and placed it in his pocket. 'Ah know things is hard for young people and families living on de Line but we will try to solve these problems. Tomorrow some of my workers will come and build some box drains and I will talk to WASA to install two more standpipes. Also when we win we plan to have a Commission of Enquiry into the coup.'

'Coup?' Harriet paused and thought about the words she would use in making her request. 'Ah doh care about no coup. Dat 'appen since 1990 and is history and we cannot change dat. All dem commission is ah set of ole talk. Ah really want yuh to geh a wuk for Justin. Leh him learn a lil trade. He have plenty talent too bad. In de morning he does geh up early and bounce basketball den in de evening he kicking football in de savannah. If all yuh could build a basketball court and put up some goalposts in de savannah.'

Carol nodded. 'When I win, I will take these suggestions to the County Councillor for the area and see if something could be done for Justin and others his age. Marm do you have a husband?'

'Yes, buh he in hospital today. He came home late larse night and was drunk and we fight and ah beat him up. Dis foolishness have to stop.'

Justin and Malcolm were embarrassed.

'Remember, we will rise,' said Imelda.

Malcolm received a text message on his BlackBerry phone. It was a message from Marcus stating that Kyle planned to return home next week. After reading the text he had a broad smile on his face. 'Well everyone thanks for coming. Everyone in dis house voting for Power to the People Party. We have to save we country from de odder party.' He was disappointed with the promises and rhetoric. He wanted to hear about the party's plans to create employment and reduce crime.

The election group of three persons moved to the neighbour's house.

Harriet looked at Malcolm. 'I voting for de government and you should do the de same. Dem people want to fool we. Look de odder partee already give we five hundred dollars to vote for dem. Remember when we were younger we did hear a story of a politician who shoot his own car to geh sympathy from de public but he still lorse.'

'Yeah ah remember dat. Marcus just send me a text saying dat Kyle coming next week.'

Harriet was overjoyed. 'Great! He is my favourite nephew.' She had a worried look. 'Yuh tink people go lissen to de polls dat predicting de new party will win?'

'Na doh worry. Two psychics already say de prime minister go win again.'

The doorbell rang and Harriet hurried to the door. She wondered if more politicians were in the neighbourhood.

An unkempt lady with a plastic bag stood at the doorway.

'Do you have anything today?'

'We eh have nothing today.' Harriet began to close the door. 'People does feel dis is de Salvation Army.'

The beggar shouted, 'Yuh take meh United States dollars and now yuh want to chase meh away!'

Harriet was shocked when she heard the accusation.

'Me?' She turned and shouted to Malcolm. 'Ent you does help she? See how she reward we?'

Malcolm quickly came to her rescue. He closed the door.

The beggar lady cursed and departed for the neighbour's home.

'Harriet doh worry she gone.'

Harriet went upstairs. She felt dizzy and knew her blood pressure must not rise and decided she needed to lie down and relax.

Malcolm remained downstairs and turned the radio dial to 101 FM. The deejay, Balliram, was accepting requests from callers. The deejay was a Grenadian who spoke in a false United States accent. He felt the accent would encourage persons to accept and agree with his nonsensical statements and biased viewpoints.

A female telephoned the deejay and said, 'Ah calling to big up meh man.'

Another caller shouted, 'Ah want to give a shout out to meh uncle and aunt from Penal.'

Malcolm steupsed and switched radio stations. There was a news bulletin warning citizens of an approaching tropical storm that was going to pass over Trinidad between 7 pm and 2 am. He wondered if he should still attend the party at his cousin's home in Marabella. He dialled Ritchie's number. 'Yeah is me, all yuh still having de partee tonight?'

Ritchie answered in a sleepy voice, 'Yeah of course. Today is meh birthday. Ah cyah postpone a partee! Storm or no storm…the partee is on. If de storm geh serious we will go for safety by de tsunami shelter de prime minister building at Tarouba.'

Malcolm quickly showered and went to his room. He opened his cupboard and selected clothes for the party. He sprayed Drakkar cologne on his chest and applied a thick layer of gel to his hair. He took the toothpaste and placed a thin strip on his toothbrush.

Jessika and Delisa also planned to attend Ritchie's party. Malcolm agreed to transport both women to the party. Jessika and Delisa arrived at five o'clock and sat in the porch of Malcolm's home. Both began chatting.

'Jess dis is ah new tank top from de flea market and if yuh see how young ah looking in it.' Delisa pointed to her stomach. 'Got two new belly button rings and three tattoos on meh waist. Going to a fete involve more dan jus' buying a ticket. Yuh have to have accessories- fix yuh nails, go by de hairdresser, buy new clothes, go for a facial, buy bracelets, high heels, lipstick and earrings.'

'Yes gyul. Larse week ah bought ah tight, tight pants, it take meh one hour to put it on and is for the Outrageous in Red concert and ah will keep it on till next week's UWI fete. Gyul, yuh going to any fetes this month?'

Delisa patted her belly. 'Nah. Ah eight months pregnant, de chile due anytime now. Also Aaron writing SEA exams in a few weeks and he sisters doing CAPE and CXC. Ah want to stay home and make sure dey study and doh idle.'

'Dat is small ting, Aaron is ah big man and he sisters is big 'oman leh them handle dem stories. Doh leh dem tings disrupt yuh social life. Remember is only one life we have to live. So live life to the fullest. Come to WASA fete, Rupee yuh boy go be dere and he going to mash up de place with Iwer.'

Delisa sighed. 'Yuh right. Fete more important. Aaron go probably never amount to anything in life…He growing up jus' like he no good fadder. De fruit doh fall far from de tree. And meh two daughters already have boyfren who go take care of dem.'

'Of course ah right. If everybody tink like you den all dem concert promoters and artistes will starve for de Carnival season.' Jessika nodded. 'So far, how much of dem parties and all-inclusive fetes yuh miss?'

'Only two ah miss. So far ah only gone to de Moka and Kama Sutra fetes.'

Delisa's cellphone began ringing. She opened it and began talking. The caller was her husband who requested a bottle of shampoo and conditioner. She would carry the items on her next visit to the prison. She closed the phone and continued the conversation.

Jessika became excited. 'In de Moka fete I pick up one sweetman, he tall like a NBA basketballer, dark like Cadbury chocolate and extra handsome.…like Shah Rukh Khan and Usher.'

Two months later there was a change in government when Power to the People Party won political power. It was a political storm which many pollsters predicted. The psychics were wrong but this did not affect their lucrative profession.

Despite the change of government, social problems continued to persist in society. Ritchie read the front page of the *Newsday*. He was frightened to see that there were three gruesome murders during the weekend. The murder toll for the year was five hundred and six persons. He was afraid of being murdered and felt the safest place would be in jail. He decided to commit the minor crime of shoplifting so he could be convicted for a short period.

The judge queried the reason for Ritchie's decision to shoplift.

'Well your honour ah doh want to waste the court time. Just give me ah jail sentence. Santa…ah mean Satan…make me do it. Ah usually doh do dat sort of ting. I usually good and does try meh best to behave buh larse week Satan was too strong.'

Some persons laughed and the judge cautioned them on their outburst and told the court that this is a serious offence. Ritchie was sentenced to sixteen months in prison.

His mother regularly visited but after four months was diagnosed with cancer. She began chemotheraphy and had to abruptly curtail her visits to the prison. She felt Ritchie would worry and decided not to inform him of her physical condition.

After one year Ritchie was released. He returned home but it was empty and lonely. Neighbours did not speak to him. The non-governmental organisation, Vision on Mission, was assisting him in finding a suitable job. Two days after his release he was visited by two close friends- Terrance and Mabel.

Terrance had a colourful bandana tied around his head. 'Some of de boys hear yuh come out today, yuh want to celebrate with a drink?' He moved his mouth sideways and rapidly blinked his left eye.

Ritchie shook his head. 'Nah, not me and alcohol again. Ah learn a lot in jail and is now a member of Alcoholics Anonymous.'

Mabel was confused. 'What?' She scratched her stomach. 'What about a smoke?' She offered him a cigarette.

'Nah, my body is a temple. I now keeping it clean and pure.'

Terrance asked, 'Yuh want a joint of ganja?' His left eye and shoulders twitched. He had a tic which developed after two years of being a cocaine addict.

'No, ah staying off all dem vices,' responded Ritchie. 'Ent ah jus' tell all yuh meh body is ah temple.'

Mabel coughed and spat on the ground. 'Well at least come and say hello to de rest of de gang from de ghetto.' She coughed and patted her chest.

'Lissen good, I going to find new and dependable friends.' Ritchie had a stern voice. 'When I was in court and had to sell meh things to pay for lawyer all yuh never help. When ah was in jail yuh never write or visit. Meh only friend was a man who get his doctorate in five months from an online United States university.'

'What is a doctorate? So how was it in jail?' asked Terrance.

'Ah doh know and doh care what is a doctorate buh de prisoners and prison guards did respect him. Jail not nice at all. Jail wasn't built to ripe figs. Every night I used to hear big men bawling and screaming.'

Terrance smiled. 'Why? Dinner was cold?' He was being sarcastic.

'De prisoners were being raped by prison officers and other prisoners. Some of dem geh HIV and dead in jail.' Ritchie was quiet for a few moments. He thought about his father who recently died as a result of AIDS and a tumour. Whilst in jail he witnessed the painful suffering and slow death of two inmates infected with HIV. 'In jail de food is a piece of stale bread and cheese or soup is cold water with a sick dumpling and a half dead piece of potato.'

'And for dessert? Wha' 'bout television and sports?' asked Mabel. She was eager to learn of the recreation and amusement in prison. 'Ah hear dey does plant garden.'

'Dessert? Yuh have a choice of catching cockroaches or rats. Chasing after them is part of the sports programme at the jail. Some days ah didn't bathe, because it hardly have water and no soap and ah did feel shame bathing in front two hundred other prisoners. I had to share a toothbrush with nine men. And we often had no toothpaste. Television? Well yuh have a choice of watching the floor or the walls. We have to plant garden or else we go starve.'

Terrance detected the sarcasm and feigned surprise. 'What? No privacy? Why yuh never complain?'

'Privacy? Boy when yuh in prison dey give you a towel, piece of soap and a bucket. Dat bucket is yuh toilet. And when yuh in dere, yuh leave shame outside. Who we go complain to? Prison is not a hotel. It made for we to suffer.'

Mabel's eyes reflected surprise. 'I shocked. Ah always thought it was comfortable like de Hilton, Marriott, Holiday Inn or Sheraton Hotel and dat going to jail was like a relaxing holiday in Disneyland. Ah want yuh tell meh six chirren all dis so dey go behave demselves and not join any gangs in de community or bad company in school.'

'Hotels? You crazy! De only similarity is dat Disneyland have Mickey Mouse and in jail have real rats, some of de rats like cocaine. All yuh real ungrateful not to visit.'

She coughed. 'Okay, so dat is wot dis is all about- we not visiting you.' She steupsed. 'If you want to move dread. No problem.' She rapidly blinked her eyes. 'Come Terrance leh we ride out.'

Both hurriedly left the premises. Ritchie was disgusted with his uncaring friends.

A few minutes later there was a knock at the door and he opened it. He thought it was Terrance and Mabel who had returned to try and convince him to visit his former gang.

An old woman stood outside. 'Morning, morning. You is Mala son?'

'Yeah why?' asked Ritchie.

'Yuh have to visit yuh mudder who in Sangre Grande hospital. She has breast cancer and de doctor say she only have a few days to live.' The old woman turned and walked away.

Ritchie looked at the dark skies and lightning. The radio and television issued a storm warning. He decided to make the trip to the hospital. Upon reaching the hospital, he found a vacant spot in the car park. As he entered the hospital, his cellphone rang. A neighbour called to inform him that his house was on fire. Neighbours were assisting with garden hoses but the fire blazed uncontrollably.

Ritchie immediately headed for his car. He drove through the heavy downpour and wanted to quickly reach his home. Ten minutes later he entered his neigbourhood and saw smoke emanating from his home. There were neighbours, policemen, newspaper reporters and television crew but no firemen. After four hours the fire truck eventually arrived but proved to be useless because the hydrant had no water.

The storm helped extinguish most of the fire. One room was not destroyed. This was a prayer room that had a shelf with inspirational and religious books including the Koran, Bhagavad Gita and Ramayana. Ritchie looked at the black clouds of smoke rising into the sky. Mala died that night and Sunita was by her bedside.

Kyle was excited. He had completed his degree and returned to Trinidad and Tobago. Marcus was glad that his son had returned home in June 2010 and was able to view the televised coverage of the World Cup in South Africa. There were concerns from some Africans that the World Cup was being staged in South Africa. The heavy sponsorship and advertisements by alcohol and cigarette companies only served to encourage vices as smoking. Also, some

Africans wondered if the money being spent by South Africa in the World Cup could have been better used to assist the poverty-stricken, unemployed citizens and to provide better medical services. These concerned and rational voices did not stop the world's most loved game from occurring in South Africa.

Every Saturday, Marcus would visit the fisherman at the market. One morning Carla accompanied her father to the fisherman. There was a traffic jam due to the police stopping cars and asking drivers to remove the tint from their windows.

Marcus said, 'Dis country really badly off. Heavy tinting illegal yet all dem car shops still placing dark tint on the windows. Is de car shops who selling it, dem wrong and should be charged by the police!'

'Is de same ting for fireworks. Stores still selling it but it illegal to set dem off,' said Carla. 'Dis country real upside down!'

Marcus stuck his head out of the car window and shouted, 'Aye wha' yuh ketch dis morning?'

'Come and see. Ah have Cavalli, Moonshine, Grouper, Cascadura, Red Snapper, Blue Marlin and some shrimps.' He waved a cloth over the fishes to chase away the flies.

Marcus parked the car near a No Parking sign and exited the vehicle. He walked to the fish vendor and examined the day's catch. 'Yeah give meh two Blue Marlin, three Moonshine and four pounds of shrimps.' He enjoyed shrimps but did not like cleaning them.

The vendor weighed the shrimps on his scale then placed them in a plastic bag. 'Anything else?'

'Nah dat is enough. In meh freezer ah still have some salmon and shark ah bought larse week from you. You ketch dis or yuh buy it from somebody else?'

'Ah does ketch it….wake up early and head to Icacos before sunrise. De Coast Guards is ah waste of time. Dey never protect we from de Venezuelans and drug lords who does shoot at us, seize we boats and cut we nets.'

'Buh maybe Venezuelans do dat cause all yuh in dem territory.' Marcus was trying to be objective. He opened his wallet and withdrew some money.

'Not at all. You ever see fish speaking English and fish speaking Spanish? Fish is fish and we know de boundaries.'

'Oh. Okay den.' Marcus paid him and placed the bags in the car's trunk.

They stopped at another stall and purchased sappodilla, caimate, soursop, pommerac and pommecythere.

When Marcus arrived home, he began to clean the fishes. He removed the fins and used a steel brush to remove the scales. Carla stood nearby and watched her father. It was an event which she enjoyed watching during the Saturday mornings. Marcus cut open the fishes and removed the guts. The fishes were cut into slices, washed and placed in a basin. She took the basin and the bag of shrimps to the kitchen.

Marcus added local seasoning to the fishes. He licked his lips. 'We go fry dis in de evening.' He placed the shrimps in the freezer. 'Fish is brain food. Everybody only have good tings to say 'bout fish and health.'

'Buh some researchers and scientists discover dat pollution in de sea like lead and mercury are stored in the fishes,' said Kyle.

Marcus was not aware of these findings. He looked at his son. 'Yeah ah sometimes read dat buh dese days everyting have pollution. Look even chicken have ah set of steroids and antibiotics. And some of de vegetable and fruits have pesticides. No wonder people getting cancer and sick at a younger age. If we food not polluted den it have preservatives or colouring.'

Near the television was a black and white photograph of Bob Marley playing football. He was dressed in a jogging suit. Kyle sat on the couch and casually flipped channels. On Channel 4, he saw a rerun of a game between Police FC and Joe Public. Both were teams in the country's Pro League division. Jean Rochford, a central midfielder of Joe Public, received a superb cross from a central defender and looped a header into the far top corner of the net. Joe Public's goalkeeper stood calm between the uprights. He was a symbol of confidence and stability. His two saves in the first half was an indication of good positioning and reflexes. In the final minute a freekick, outside the penalty box, was given to Joe Public. Jean kicked the ball over the wall of four defenders. It curled towards the left goalpost but was headed away by a vigilant defender.

Kyle yawned and decided to sleep. He switched off the television and headed for his bed. Later, in the night, he and two friends attended the International Soca Monarch competition. He was horrified to see the low quality of performances.

The next morning he descended the stairs and met his father preparing breakfast. 'Dad did you stay up to see the competition?'

'Ah saw a few performances. Dem shows is ah waste of time.' He placed a plate with fried eggs and bread on the table. 'Here eat this.'

'In the past I wasted my time and money. Not me again. All the artistes had scantily-clad women wining down de place. A total absence of choreography in the dancing. It certainly looked vulgar. Loud music and lyrics I could not understand seem to be the crowd favourites.'

Marcus placed a tea bag in an old, enamel cup filled with boiling water. He added a spoonful of brown sugar and stirred with a silver spoon. He added condensed milk and stirred. Kyle carefully poured orange juice into a glass.

'Son, people today doh know 'bout de history of carnival and calypso as forms of resistance. De artform has degenerated and today we accept it as part of our culture. See de thousands of people who attended larse night and went home early dis morning. Dey like dat and once de audience like garbage… den de artistes will continue to produce a wide selection of garbage. Look at the Soca Chutney competitions with all dem songs on drinking rum and horning. During indentureship days, it had songs dat was powerful in meaning and a form of resistance for de labourers. Today all kinda dotish songs winning first prize of two million dollars. For the Calypso finals it have some calypsonians who singing racist songs or with shallow lyrics and dey too winning de big first prizes.'

'How could de government give two million dollars for these first prizes? Could our country's treasury support this? Why the organisers and promoters referring to the competitions as 'International' when most of the artistes are locals? I was at the Savannah a few years ago and shocked to see the small costumes that cost so much. Is like that story -The Emperor's New Clothes. All dem masqueraders parading half-naked and fat people saying dey looking so good and saying dey having a good time.'

Marcus stopped chewing and drank tea. 'My voice and your voice mean nothing. Dem singers will still be singing wave yuh rag or flag and move to the left and move to de right. Band leaders will continue making small costumes for people with a small brain. Imagine in June and July people launching Carnival bands. What hurts me is to see innocent children doing vulgar wining. De government has to spend big money to geh votes and to prevent people from saying dey racist and doh like culture. If de government has to borrow from de IMF and World Bank to pay dese big prizes…den dey will do it.'

'Buh dat is not culture. That garbage is not even pseudo-culture. That is something superficial and shallow masquerading as culture. Larse year, Delisa was drunk and was wining on fire hydrants and all de lamp posts on St. Vincent Street. She thought it was midgets and tall tourists. We no longer scraping the bottom of the barrel, we now digging under the barrel! It is this display of shallowness which makes us a backward banana republic. Ah cannot understand how de government could spend millions of dollars on Carnival while so many people below de poverty line and unemployed.'

Marcus had changed. He was always patriotic and praising the culture of Trinidad and Tobago. However, during the past four years the feelings of nationalism had subsided. He looked at Kyle. He was glad to see that his son was able to critically assess the level of cultural development. He took two slices of bread and placed them in the toaster oven. 'Son forget about dem. At the end of de day is me and you to ketch we tail. Focus on getting married, find a good job and build or buy a house. Doh do like some citizens who go on tv and cry long tears and beg de government for a house or handouts.'

'Oh, larse night ah saw Mister Bally wining down de place with two fat women. He big in de dance.'

'What? Bally dat slacker still alive! Long time ah didn't see him.' Marcus remembered the colourful character who was always chasing after women.

Kyle removed the bread from the toaster and placed butter on both slices. 'Yeah. He lorse a lot of hair and have a paunch. He tell me that he planning to go and live for while in the Middle East.'

'Middle East? Where? I heard some people going Dubai to work.'

Kyle had a serious expression. 'He planning to join ISIS.'

Marcus steupsed and laughed. 'Bally never change. He is one arrogant fella. Always living in la-la land chasing after imaginary enemies and saving people. He like to feel he is a James Bond.'

Kyle smiled and was relieved. He left his father and went into the living room. He saw a football magazine and began flipping the pages. He planned to attend the FIFA Under-17 Women's World Cup during 5 to 25 September 2010. Trinidad and Tobago had automatically qualified as it was the host country. He remembered viewing, via cable television, the Under-17 tournament in New Zealand. It was an exciting final in which North Korea defeated United States 2-1. A month later, he eagerly watched the televised draw ceremony at the Hyatt Regency Hotel in Port-of-Spain. He read the reports on the training and performance of the 'Soca Princesses' who would represent Trinidad and Tobago at the Women's World Cup. His country was in Group A which had strong competitors as Chile, Nigeria and North Korea. He was also interested in the performance of Germany in Group B and Brazil in Group D. These were his two favourite football nations. He felt the prices of tickets were reasonable. It was forty dollars for the group phase and sixty dollars for the knockout games. He viewed games at the Larry Gomes Stadium in Arima and the Manny Ramjohn Stadium in Marabella. He had accompanied his father to these venues in 2001 to view the FIFA Under-17 Men's World Cup. One of the memorable goals was a simple chip of the ball, by a midfielder, over the advancing goalkeeper and into the net.

Kyle slowly entered the turnstile at the Larry Gomes Stadium. He wore a chequered shirt and a blue track pants. In twenty minutes there would be a tense quarterfinal encounter between Japan and Ireland. He supported the underdog- Ireland. The Japanese girls were accustomed to short passes whilst the Irish used long passes. Ireland's weakness was their lack of aggressiveness and they seemed comfortable with back passes. Kumi Yokoyama dribbled past two defenders and sent a curling shot into the top corner of the goal. The crowd responded with light clapping. Then Megan Bell, of Ireland, fired a volley at Japan's goalkeeper but it was not a threat as the ball went over the top bar. He felt Bell would soon score since she had converted a forty yard free kick into a goal in a previous game against Ghana. The final score was 2-1 in favour of Japan. On the way home, he turned the dial of the car's radio. He heard that Spain had defeated Brazil and the game had a similar outcome 2-1.

Marcus pointed to a large crowd at the rumshop. It seemed that the global recession did not affect bars and rumshops. Tomorrow he would view the highlights on the television. 'So do you think that more people watch Super Bowl or the World Cup?'

Kyle immediately responded. 'De World Cup of course. De Americans want us to believe dat everybody does play and watch dat kinda football. Think 'bout dis- how many Asians play American football? How many people in Africa play American football?' Whilst in Canada he heard disparaging remarks of Americans and the manner in which they portrayed their sports to the rest of the world.

'Yuh right more people like dis type of football. Yuh hear dat Beckham coming here next week?'

Kyle was excited when he learned that David Beckham would be visiting Trinidad. He wondered if Beckham regretted leaving Britain to play football in the USA.

Beckham launched a football festival at the Marvin Lee Stadium at Macoya and also planned to attend the finals of the Under-17 Women's World Cup at the Hasely Crawford Stadium.

'Do you know about the Christmas Truce of World War One?' asked Kyle.

'No. Never heard 'bout it.'

'It was a time on Christmas Eve and Christmas Day on the Western Front in Europe when British and German soldiers stopped fighting and were friendly towards each other. Some of the soldiers exchanged food and tokens. And the British, French and German soldiers played football.'

Marcus smiled. 'See the power of football to create peace amidst war.' He parked the car and closed the garage gate.

Next day, Kyle and Marcus arrived early at the Hasely Crawford Stadium. It was the finals between South Korea and Japan and also the third-place playoff between North Korea and Spain. Marcus spotted Francine in the crowd, 'Aye Fran come na gyul.'

They hugged each other. She was his former girlfriend.

Francine hugged Kyle. 'Kyle yuh well get big.' She patted his head.

Kyle nodded. He did not like her.

She glanced at Marcus and had a twinkle in her eyes. 'Love when yuh coming to check me?'

'Fran these days ah busy…buh soon we go link up.'

'Ah understand. Ah making a trip to Grenada and den New York soon and if yuh free, yuh welcome to come along to New York.'

He smiled. 'Yeah for sure. Will talk later. Bye.' They lovingly embraced.

'Bye honey, I really miss you,' she whispered in his ear. She kissed his cheek then quickly disappeared among the crowd of football enthusiasts.

After he and his son went searching for a seat. They found two empty seats. Marcus was quiet. He was thinking about Francine. Kyle began turning pages of the *Newsday*. There were two photographs of a football game, in Haiti, with players who were amputees. The players had lost their limbs on 12 January 2010 when a massive earthquake killed more than 220,000 persons in Haiti. Indeed, football was a game that could be played and enjoyed by the physically and mentally challenged.

Marcus made a bet with Kyle that North Korea would be victorious. Spain scored first with a goal that slipped past the goalkeeper's hands. The crowd hoped for an equaliser from North Korea to make the game more exciting. It never happened. Spain dominated the game and eventually won the third spot with the final score being 1-0. This was a different level of football.

Marcus realized that in the Under-17 games there was an absence of popular names such as Samuel Eto'o of Cameroon, Lionel Messi of Argentina or Cristiano Ronaldo of Portugal. 'See how football cross ideology lines. Nobody here today saying North Korea is communist and because of dat dey should lose. It was de same ting when there was East Germany and West Germany. Fans support a country if the players have skill.'

Kyle nodded.

The final game of the Under-17 World Cup began with the shrill blast of the referee's whistle. The captain of South Korea's team, Kim Areum, seemed confident. She ran into the 18 yard box of Japan and fired a powerful shot. The South Korean goalkeeper made a great reflex save. The crowd cheered. In a quick comeback, Japan's Hikaru Naomoto fired a long range shot. The ball slipped through the gloves of the South Korean goalkeeper and rolled into the back of the net. Kyle looked at his watch. Only eleven minutes had elapsed and the score was 1-1. Kim Areum of South Korea ran towards the goal. She received a short pass and tapped the ball into an open goal. At the end of ninety minutes the score was 3-3. The twenty-two players on the field were tense. This was the game of their life. For some it would be the moment of glory in their life. After extra time the score remained unchanged. There would be penalty kicks to decide the winner. The faces of both captains reflected concern and anguish. They had to choose their best to take the penalties. There must be no mistakes. South Korea defeated Japan 5-4 in the penalty shootout. The exciting Asian battle had ended.

Both Marcus and Kyle decided to leave and not witness the prize-giving ceremony. Marcus wanted to avoid the traffic which occured after such events.

'This is an exciting finish to this World Cup game,' said Kyle.

Marcus descended the stairs. 'Yep. Dese spectators got dere money worth. Yuh grandpa would have liked this game. He was here to witness dat qualifying game when we played de USA.'

Kyle had heard of the historic game. 'Yes, in 1989.'

September 2010 was a month for football. The Under-20 Soca Warriors were in Guyana to participate in the Kashif and Shanghai inaugural President's Cup. The Soca Warriors were to play Guyana at the Guyana National Stadium. Kyle was not interested in their games but was glad to read the newspaper report that his country's team had defeated Antigua and Barbuda 1-0 and St. Lucia 3-0. Marcus too showed little enthusiasm in regional football such as the Digicel Caribbean Cup and the CONCACAF Gold Cup.

In 2011, Kyle followed the Champions League in Europe. He supported Lille which was at the top of the French League. He checked ESPN and saw one of the games. He hoped Lyon would win in their game against Toulouse. Lisandro Lopez unleashed a shot to the right top corner but goalkeeper Pierrick Cross dived and punched the ball past the right post. Marcus told him that some countries such as France, Italy, Brazil and England had certain unique styles of football. Kyle was a keen supporter of FC Porto, Benfica and Braga and glad to see that the Portuguese clubs had qualified for the Europa League semifinals. He was surprised to learn that Spartak Moscow had not been able to qualify for the League.

During lunch, Marcus was reading the sport section of the newspaper. He read that Brian Lara and Dwight Yorke would be playing in the British Airways Football Legends Invitational at the Kensington Oval in Barbados. He turned to the page with local sports. 'This evening is a National Youth League game at Waterloo Secondary School ground. Yuh want to go?'

'Who playing?' asked Kyle.

'Club Sando and Cap Off Youths.'

'Yeah. Next week we should also go to see Group A games with FC Santa Rosa, Mayaro Spurs, QPCC and Eagles United. Ah hear de standard higher.'

'Go and ask Carla if she want to go dis evening. Call Justin and see if he interested.'

Kyle nodded and left the kitchen. He headed upstairs. Carla agreed to attend the game. He called Justin's home but nobody answered. He descended the stairs and returned to the kitchen.

'Yeah she going buh nobody answering by Justin.'

'Okay, we leaving in the next two hours. Did you read dis article about a Trini who playing for Orlando City in the American Division of the United Soccer League Pro Championship?'

'No.'

'Yeah, he score a goal yesterday against Richmond Kickers at the Citrus Bowl in Florida. It have West Indians all over de world playing football or cricket.' Marcus gave the newspaper to Kyle. 'And there is an article on de next page 'bout de Digicel Kick Starts clinics and its Academy. So much financial support being given to football and so little results.'

Kyle wanted his community to participate in the Guinness Street Football competition and also the five-a-side football competition to be held at the Beach Volleyball Courts at Saith Park in Chaguanas.

His father suggested that he ask a welder to make two small goalposts. 'Check outside there might be some iron to make goalposts.'

Kyle nodded and left. He found a few scraps of iron and steel in the junk room. He brought them for his father who was watching ESPN. Marcus was a keen follower of Futsal and hoped to attend the upcoming World Championships. 'Yuh know ah find it strange that futsal competitions never catch on here. Plenty people does play on de street and savannah. We doh appreciate local football. Not many know 'bout de Port-of-Spain and Environs Boys Football Championship. On Friday is ah big clash between Carenage Boys Government and Eastern Boys Government at Fatima College grounds buh not many people know 'bout it.'

'That's true and not many people know that there is a Girl's Division with St. Agnes RC being the top team. Look how poorly attended that final with

WASA FC and Maloney FC was in the Eastern Football Association. The Marvin Lee Stadium wasn't even half-filled.'

'Something like futsal which is played on the streets in Uruguay and Brazil should have spread here in a big way. One of meh good friends, Ming went to the World Championships in Las Vegas a few years ago. Lissen, dey having some football at UWI next week. Check de dates and we could go to a few.'

Kyle nodded.

Marcus and Kyle attended three of the games of the University of the West Indies Super Football League. Marcus supported Titans and House of Dread whilst Kyle's favourite was Star House. He met Aboud who had a son playing for Athletico. These teams were in Group B. Kyle enjoyed the Group A game between Caution and Athletico. It ended 6-2 in favour of Caution.

Next morning Marcus checked his emails and saw two messages from Francine. He was hesitant to read them and decided to check the results of the match between Club Sando and Real Maracas in the Blink/bmobile National Super League held at the Mannie Ramjohn Stadium in Marabella. Club Sando won 1-0. He hoped to attend other matches in the League involving Stokely Vale, Eagles United and West Side Super Starz. He had cousins who were playing for Brick Field Falcons and Los Iros in the Roy Jagroopsingh Second Division. At the end of the month, he and Kyle would attend the Executive Cup quarterfinals between Siparia Spurs and Giants at Guaracara Park in Pointe-a-Pierre. Both teams belonged to the Petrotrin Super League.

He checked the calendar on the refrigerator. There were scribbled notes on some days. Tomorrow would be the start of the Eddie Hart League Football Premier Division Big 8 Competition. The first game at the Eddie Hart Grounds in Tacarigua was a double header with EM Madrid Pacemakers and Trincity Nationals. Two complimentary tickets were attached to top of the calendar. These were for the Toyota Classic Cup finals at the Larry Gomes Stadium in Malabar. Football season was a busy time for Marcus. He felt it was more hectic than Christmas.

In August 2011, the government declared a State of Emergency and enforced a curfew to deal with the high level of serious crimes. Kyle attended two curfew parties at the homes of friends in Cunaripo and Sangre Grande.

One evening after the television news, Kyle told his father, 'Imagine the Soca Warriors since 2006 waiting for money dat de Football Federation owing dem.'

Marcus nodded. 'I'm not surprised. Dat seems to be normal in Trinidad and Tobago. Yesterday I read that since 1991 residents in Oropune and Piarco waiting for compensation due to their relocation. In another country this long wait would not have been tolerated. Speaking of dem footballers, de West Indian Players' Association planning a fun game called Balls of Fire between de Warriors and some cricket players.'

'Yeah, read about that. Sometimes I wonder 'bout how money being spent in dis upside down country. Imagine the coup enquiry costing more than the damage and destruction that 'appen in 1990.' He paused. 'Earlier today I saw BBC news reporting on rioting in London. It started as a protest against police shooting a man from Tottenham.'

Marcus was pensive. He remembered his father's conversation about two policemen who killed his grandfather. He was glad to see that his son was educated and admired his independent thinking. Kyle noticed the sudden silence and decided to change the topic.

In February 2018, Marcus took Kyle to see the Black Panther movie. Both men were not impressed and during the movie were bored. Marcus said, 'Son these movies are overrated. The plot is weak.' He looked to see his son's reaction. In the nearby rows, some moviegoers were dressed in ethnic wear including dashikis and head wraps.

His son smiled. 'Yep. We talked about the usual hype and exaggerated reviews.'

As they exited the cinema, Marcus asked, 'Why do you feel this movie making so much money and everybody in United States and other parts of the world talking about it?'

'Ah not sure…blacks especially Afro Americans need a hero. Some of their heroes have failed them. Look at the racial profiling and police killings of unarmed Blacks. Not even their country's first black president could save them from that racism. And it's even worse now with Trump and his rhetoric.'

'Exactly!' The father nodded. He felt proud of his son's analysis and insight.

'Blacks don't need fictional heroes. We have real people in the past such as dem Africans who resisted slavery. Dem Americans have men like Booker T. Washington, W.E.B. DuBois and Blacks in Africa have builders of temples and pyramids in Ancient Egypt.'

'Agreed. Not just the past but so many talented people alive today.' Marcus felt that his son had matured. They finally reached home. Marcus closed the car and Kyle headed for the front door.

Kyle blurted out, 'So let's go to the next World Cup.'

His father was pensive. He listened to the neighbour's dogs barking that drowned the peaceful chirping of crickets in the nearby bushes. 'Not sure if we can afford it but start saving your money. Maybe for de World Cup in 2026…we could go to it.'

'I've already started to save money for both of us to attend. And I hope Trinidad and Tobago going too!'

In December 2019 residents of Jordan Street had not heard any reports of the mysterious virus that originated in Wuhan, China. One month later, they were unaware that the president of their country had issued a proclamation about COVID-19. The neighbouring residents of Vierra Street were also unaware of the impending global pandemic. Both streets were located in Enterprise. These were dangerous times in the world but for the residents in these streets it was normal.

Aaron and Mabel lived in Vierra Street. Aaron was fifteen years younger than Mabel who was sixty years old. They had nine children. Aaron had three children from a previous relationship. One of his sons was autistic. Mabel had six children from two earlier marriages. She wore a jersey with the words- 'I survived Ebola, SARS and H1N1'.

Aaron looked at Mabel, 'I'm so glad that we have united and have a large family.'

Mabel steupsed and rolled her eyes. 'Why?' She always felt it was difficult for them to provide food for all the children. Also she felt cramped resided in the small, one storey house that had three bedrooms and one toilet. The walls were unpainted and during windy days the galvanize that comprised the roof would be slamming against the wooden beams.

He smiled. 'Don't bother, forget I brought it up.'

In the past year, Mabel had three extramarital affairs. She felt it was a generation curse because her mother and grandmother were also unfaithful to their spouses.

Aaron knew of the relationships but chose to ignore her unfaithfulness. His friends advised him to leave her and move on with his life. He always smiled and responded his friends. His answer was the same, 'I not bothered about that. Yuh know I always wanted a short, sweet stocky woman. She will soon stop that.'

Despite his casual dismissal of the situation the embarrassment affected Aaron. In December 2019 he attempted suicide by ingesting a pesticide but his eldest son saved him. Mabel participated in one of the medium size bands on Carnival Monday. Next day she experienced acute diarrhoea. Aaron felt it was due to eating spoilt food. The diarrhoea lasted for three days. On the fourth day, Mabel collapsed and died. Nobody realized it was due to the coronavirus.

At the funeral service, Cleopatra whispered to Aaron, 'I heard diarrhoea was one of the symptoms of COVID-19.' She was saddened and felt weak.

Aaron was glum and his emotions were mixed. He felt grieved but also relieved as he would no longer be embarrassed.

Bamboo Foot lived opposite the unkempt park on Vierra Street. Twenty years ago he killed a man who was cheating in a card game. He was sentenced to life imprisonment. In 2014, he received a presidential pardon and decided to rent a room on Vierra Street. The news of COVID-19 made him panic. He saw neighbours buying groceries and became worried.

He decided to go to the groceries and purchased three hundred rolls of toilet paper and four hundred cases of bottled water. At Price Smart he saw Cleopatra, Harriet, Harrilal with trolleys packed with groceries.

At end of February, Bamboo Foot died of pneumonia at the Caura Hospital. He had contracted the virus from an infected shopper at the grocery. He was the country's second COVID-19 death. Interestingly, the population was focused on those who were infected and died from the virus. Nobody cared about the murder statistics which was 167 deaths and also persons who died from cancer, diabetes.

In March, Baldeo, during the Crime Watchers, he coughed and said, 'Whoa like ah getting de coronavirus. Dem gyul and dem want to get me in quarantine with me. Live life, nobody gonna die. Ah coughing so what.' He laughed and brushed it off believing it was a dry cough. The coughing continued for two days and he decided to go for a test. Two days later, a video began circulating with Baldeo. He was crying and speaking to his daughter. He said, 'Gyul I going to die I have de virus.' Then he took a sign from his pocket and displayed it to his daughter.

Many citizens of Trinidad and Tobago would remember 7 April 2020. It was a historic day as all KFC outlets in the country were to be closed till 30 April. The prime minster issued a statement, 'Stay home and cook allyuh food. This is a peril.'

Cleopatra shouted at the television, 'You is ah real wicked man. Is we vote for you and dis is what yuh do we. More than two weeks we have no doubles, no corn soup, no roti, no Rituals, no Subway! Dat is total madness.'

The next door neighbor, Harriet, was also listening to the message on the radio. She began to shout and curse. She dialed her common law partner. 'Lissen is me. Where yuh?'

It was difficult to hear due to the loud music. 'Ent ah tell yuh ah going for a river lime in Caura,' said Terrance. He was employed as a cashier at Popeyes in MovieTowne in Port-of-Spain.

'Pick up plenty…plenty KFC.'

'What yuh mean by plenty?' he asked.

She looked at the moving spiders on the cobwebs hanging in the corners of the ceiling. She hated spiders. 'About ten buckets of chicken, twenty packs of fries and sauce, and fifteen containers of cole slaw.'

'Ten buckets?'

'Shut yuh blasted tail and do what ah tell yuh! Doh worry about it spoiling, we will freeze some!' She was angry and switched channels on the television.

'But ah don't have dat type of money on my card and not enough cash.'

Harriet was furious. 'Boy ah don't want to cuss this hour of de day. Go to de credit union or bank and get a small loan. We also need to buy plenty Lysol and other cleaning products. Remember to also pick up de KFC Super Family Deal which is eight pieces of chicken and six regular sides. Also the Mega Meal at Royal Castle.'

She hobbled outside and shouted, 'Allyuh come inside and bathe. Just now is dinner.' She felt a tightness in her chest and began to sweat. It was a strange feeling and she muttered, 'All dis stress to order a simple meal.' For the past eight months she was employed as a waitress at the Palance Bar in El Dorado.

Her husband reluctantly agreed and hung up the phone. He poured another cup of rum. The rum made him feel relaxed and calm. He always liked a river lime with his friends. It made him forget his troubles at home and work.

The pain persisted but Harriet was not bothered. She believed only men were susceptible to heart attacks. She drank some water and sat in the rocking chair near the fan.

There were repeated public warnings about social distancing- keeping a safe distance from others. Persons on Vierra and Jordan Streets chose to ignore the warnings. Certain laws were ignored by the residents of these two streets. Moral and social codes were overlooked. These were streets that seemed to be in a different world, similar to old television series- Twilight Zone.

Harriet grabbed her back and winced. 'Oh gosh like my kidneys acting up again.' She went to get the phone and collapsed on the floor.

The residents of Vierra Street had rooms filled with toilet paper. Most of the people on Jordan Street had hand sanitizers and hundreds of cases of bottled water.

Ming was in the Golden Grove Remand Yard Prison. He was serving nine years for stealing cars. The police uncovered a plot, in 2013, in which he was stealing cars and selling them to corrupt garage owners selling car parts. Ming used his expensive Iphone to directly contact the Minister of National Security, Little Stewart.

'Hello this is Ming.'

'Good morning Mr. Ming. How did you get my phone number?' asked Little.

'A close friend of mine who knows your step-daughter.'

There was a pause. He thought about ending the call. 'Oh, ok. How can I help you? Who are you?'

'The living conditions in the Golden Grove Prison rel bad. We go appreciate better security in here, it jus' not safe anymore. Our cellphones not safe. People not feeling safe, yuh know.'

Little smiled and decided to abruptly end the conversation, 'Ok thanks I will investigate the matter and request a report. Bye.' He promptly called the telephone company and asked the manager to change his phone number.

Ming felt satisfied. Little was in deep thought. He was bothered by the recent conversation with the prisoner. He wondered if the prisoner was convicted for a serious or petty crime. He planned to tell the public that he was connected to the grassroots. He wondered why people felt he and members of the government were out of touch with reality.

Ritchie had an Apple phone and he called the Attorney General to complain of the poor quality of prison food.

In March 2020 there was a prison riot. Ming and Ritchie were involved in the protest. Ritchie was serving six months for shoplifting. The protest was due to poor conditions and fears among prisoners of COVID-19 spreading among inmates.

The inmates had improvised weapons included small pieces of wood with razor blades at the end, penknives, sharpened pieces of iron and broken toothbrushes. After the riot was quelled, the prison office found sixty cellphones, phone chargers, cigarettes, joints of marijuana and currency. Wilfred was employed with CNC3 and covered the riot.

On weekends Aaron worked at Burger King in Port-of-Spain. Due to overcrowding at the prisons, the Attorney General decided to release some of the prisoners. Ming and Ritchie were released. Ming rented a room from Aaron in Vierra Street and Ritchie lived with his aunt and uncle on Jordan Street.

For three weeks, Aaron provided food for his new tenant. Ming ate KFC and Royal Castle. 'Boss sorry we have no butter, bread and rice,' said Aaron.

'No worries man. Is better than prison food.' Ming bit into the defrosted chicken. He squeezed ketchup on the cole slaw and fries. He laughed. 'Yuh have any buns.'

'Yeah I thawing some out for you and the chirren.'

After dinner, Ming took a walk to Delisa's house. The women in both streets hated her. She worked as a cook at Pancho's on Edward Street in Port-of-Spain. She was friendly but accepted sexual favours from the men in Enterprise and other parts of Trinidad.

Delisa regularly visited New York to shop for Valentine's Day. She had a skybox and could have ordered items from anywhere in the world. However, there was a certain feeling of superiority when she boasted that she was flying abroad for the weekend to shop. Three days after returning from New York she complained of a sore throat. She went to the health centre and was tested positive for COVID-19. The doctors advised her to stay in the hospital and await transportation to a quarantine facility. She furtively slipped past the guards at the hospital and retuned to Jordan Street. She decided to self-isolate and stay at home. One week after the diagnosis she died.

Five days after Delisa's death, Ming developed a persistent cough.

A cloud of paranoia had enveloped the two streets. Ming set his alarm clock to ring every thrity minutes. This was a reminder for him to drink a cup of water. He read an email from a doctor who claimed that regular sips of water to keep your throat moist would prevent the virus from entering your body. Ming saw a report on CNN. It was the president of the USA discussing the possibility of ingesting disinfectants to eliminate the virus.

Next day, he discussed it with Aaron whose eldest son had a headache and body pains. Aaron mixed some Clorox and mopping solution with water. His son reluctantly sipped it and felt nauseated. He took a deep breath and drank the entire cup. Two minutes later he fainted. Aaron heard a thud when his body hit the floor. His son was frothing from the mouth. He began to panic. He dialled the paramedics but they did not have an ambulance. He placed his dying son into the backseat of his car and rushed to the medical centre. His son died before they reached the hospital.

Aaron cursed the government and the health system. He cursed God. He returned home and cursed Ming. 'Is allyuh damn chinee people create this! Coronavirus was created in a lab in Wuhan and dem chinee people spread it across the world to damage the world's economies.'

Ming was not sure how to respond. He wanted to offer condolences. 'Dat is not true. It was due to somebody eating a bat from the market in Wuhan.'

Aaron felt blood rushing to his head. 'WHAT? Shut yuh damn tail. Get out dis house now!'

Ming looked confused. He wanted to apologise.

'GET OUT! Get de hell out of here before I commit murder today!'

He hurriedly opened the gate and closed it. He meekly pulled the broken trolley with bags containing his clothes and utensils.

Ming, in April 2021, was tested positive for COVID-19. He decided that self-isolation was best. He did not want to use local remedies. A friend suggested the use of anti-malarial drug hydroxychloroquine. It was only available at the hospital and not for public use. He knew a nurse and she stole a large quantity. He paid two hundred dollars for a large vial of the drug. He died after taking the first dose.

In June 2021, there was another lockdown due to the pandemic. Aaron regularly boiled ginger and other herbs then drank the bitter solution. He purchased one thousand masks, six hundred bottles of Lysol and 5 boxes of face coverings.

In Port-of-Spain, there was a different scenario. Raphael was looking for food in the dustbins. Raphael did not realize that there were COVID-19 cases and deaths in the country and rest of the world. He read about the virus in old newspapers. Thomas was nearby. Both were elderly.

'I wish someone from de government would help we. Or protect we,' muttered Raphael. 'Nobody care 'bout black people. Now and den some person does give we good food but he ketching we arse.' He spent the day sleeping or roaming the streets.

Thomas cursed. He was partially blind and walked with a limp. 'We eat food off the ground, we eat food from bins, we drink anything people leave. We smoke old cigarettes. Why nobody helping we.' The closure of KFC and restaurants made it more difficult for the homeless to obtain discarded meals.

A police van stopped near the two elderly men. A policeman exited the vehicle with a glass bottle. He pointed a loaded gun at Raphael and shouted. 'You drink dis!' He took off the safety clip and his finger was on the trigger.

Raphael was frightened and snatched the bottle. He smelled it. It was pungent and seemed like alcohol. He obeyed. The police forced Thomas to do push-ups. He did two then collapsed. Another policeman ordered Raphael to run along the street. One of the officers in the van had a cellphone and recorded the incident. It was later shared on Facebook. That night, both homeless men were killed. Nobody was convicted for the crime.

The political landscape was in a deplorable state. Devon was extradited to the United States to face charges of money laundering, racketeering and wire fraud. After the PPP lost elections, Imelda felt humiliated but refused to resign as leader. Clyde departed from the PPP and began working on his autobiography. Carol had disagreements with the party's executive members and decided to leave and form a new political party.

Aaron was confused with the conflicting information on the effectiveness of vaccines. He wanted to ensure that he would be well-protected and decided to take the Pfizer, Johnson and Johnson, Moderna and Sinopharm vaccines.

This was Trinidad and Tobago with confusion and hope. The country had potential and victories were sometimes temporary and often delayed.

I want morebooks!

Buy your books fast and straightforward online - at one of world's fastest growing online book stores! Environmentally sound due to Print-on-Demand technologies.

Buy your books online at
www.morebooks.shop

Kaufen Sie Ihre Bücher schnell und unkompliziert online – auf einer der am schnellsten wachsenden Buchhandelsplattformen weltweit! Dank Print-On-Demand umwelt- und ressourcenschonend produzi ert.

Bücher schneller online kaufen
www.morebooks.shop

info@omniscriptum.com
www.omniscriptum.com

Printed by Books on Demand GmbH, Norderstedt / Germany